RACE FOR THE NET

When African Americans Controlled the
Internet and What Happens Now?

Albert E. White

First Edition 2020
Library of Congress Cataloging-in Publication Data Number TX 8-538-187
Albert. E White

Cover Designed by AMW Marketing and Design

*This book is dedicated to my parents, Luther and Thelma White,
who instilled an entrepreneurial spirit in my brother Martin
and me while promoting a love of family and friends. This
enhanced our lives by sharing their faith in God and prayer.*

TABLE OF CONTENTS

Part II: Providing Solutions and Direction for the African American Community

Part III: What Is Needed to Ensure Our Growth and Success?

Acknowledgments

I would like to thank the Beyster family for providing me additional information to assist me in writing Race for the Net. Information from their book Names, Numbers and Network Solutions — The Monetization of the Internet, was critical in better understanding the relationship between SAIC and the founders of Network Solutions. I would also like to thank the Nielsen and Pew Research Company for permitting me to utilize their research information.

Thanks to my family for their support: my wife, Gwen McLaughlin-White, my first editor and my partner; my children, Nancy, Chris, and Adrian; and my six grandchildren. Thanks to my brother, Martin and his family for being my conscience and providing mental support.

Thank you to the four founders of Network Solutions (Emmit McHenry, Tyrone Grigsby, Gary Desler, and Edward Peters) and all the other members of the Network Solutions family for making this untold story possible.

Special thanks to Dr. Lonnie Johnson, George Fraser and Thomas Hodges who inspired me to write Race for the Net.

Thanks also goes out to the following people who supported me throughout this project: Paul White, Stephanie Bates, Clyde Jackson, Ron Stroman, Wilmer and Sheila Leon, Linda Johnson, Tony Pace, Matt Lawn, Rich Stewart, Moses and Gwen Brewer, Alfreda Kemper, Lynn Jenkins, Anita Goodman, Elizabeth Olana, Virginia and Jibri Griffin, the late Ethel Griffin, Thelma Austin, Clarence Haynes, Richard and Gwen Hackney, Bob and Muriel Lawrence, June Killingsworth, George Sweeper and son, Juanita Matthews, Singleton McAllister, Calvin Vismale, Carlton Rice, Thomas Webster, Anthony Robinson of MBDELF and Warner Sessions. Thanks to John Clyburn, Marty Rosen, Stu Kerzner, Reggie Ratliff, and his sister the late Pat Ratliff.

My editing and book team: Max Fortune, Janina Lawrence, and self-publisher, Book Baby of New Jersey. A special thanks to my Kingdom Fellowship AME Church family of Silver Spring, Maryland (Sons of Joshua Boys to Men Program), Leon Anderson and his mentoring staff, Rev. Matthew Watley and his ministry staff for their prayers).

FOREWORD

Dr. George C. Fraser

In 1993, I visited an African American company in Herndon, Virginia where I met with two dynamic African American men who had been friends since 1964. They were Emmit McHenry, a former insurance executive with major Fortune 500 companies, and Albert White, a former banking executive with major money center banks in New York in the '70s and '80s. Their position as executives with major financial institutions during that time was highly unusual for African Americans. However, their positions in corporate America fit their personalities and the competitive nature they had demonstrated most of their lives. Emmit was a competitive wrestler in high school and college, and Albert was a standout basketball player in high school and college. Both had southern roots, Emmit was born in Forest City, Arkansas and raised in Oklahoma, and Albert was raised in Brooklyn, New York by parents who migrated from Prentiss, Mississippi in the '30s.

Each believed that they could achieve great things in their lives, regardless of the environment. Raised by parents with a deep appreciation of the importance of a good education, each holds advanced degrees from leading American educational institutions.

Neither one realized when they met on the campus of the University of Denver in 1964 that they would both participate in the introduction of the greatest technology of the 20th and 21st centuries, the global Internet. Emmit McHenry was the chief executive officer and Albert White, the vice president of corporate communications for the company that first introduced the Internet to the public, not just in the United States, but around the world in 1993. You will recognize the company's name—Network Solutions—since it is very active today in the Internet space, providing Internet addresses and services. The original Network Solutions was an African American owned and managed business, WOW!

Mr. White's book, *Race for the Net,* provides detailed information about this historical fact that few people are aware of. *Race for the Net* is an inspiring and inspirational book about how this African American company opened the door to the Internet that today has 4.2 billion global users, more than half of the world's population and growing. It walks us through the history of Network Solutions and its accomplishments and problems as an African American technology firm at that time. The author, Albert White, was one of the first persons to promote the Internet's global importance and use of this technology. In his chapter "Selling an Invisible Concept," he shares examples of the difficulty of this task since this technology was new to the global community and especially to the African American community. He also points out how equal access was of great concern to various communities during the early days of the Internet and how that directly relates to the net neutrality issue today.

Albert also details how the founding owners sold the company at the beginning of the Internet explosion for less than $5 million, only to see the company sold five years later and recognized as the largest sale of a technology company at that time, exceeding $20 billion. He expounds on why the founders were forced to sell the company and why the African American community nor Wall Street offered any financial assistance for them to retain ownership. The difficulty in the African American business community over the years in raising capital is not new. Even today, we are still facing a similar problem of lack of capital to maintain and grow our businesses.

In *Race for the Net*, the author shares his 30 years of business experience as an advisor to some of the most successful CEOs in this country with his unique ability to understand the direction of future technologies. With some invaluable advice to readers of all ages in areas of future growth, he points out some aspects of issues in our community that must change to ensure our participation in this fast-moving age of technology. We need to become more cohesive as a community and invest in businesses even when we do not fully understand the technology; we need to rely on the experience of those dreamers in our community.

We can no longer afford to not fully capitalize on the technological changes occurring in our society today and that will occur in the future. Our ability to support each other with capital and experiences will be important in creating job opportunities and wealth for the next generations. With things changing so rapidly today, it is important that we support each other so we do not miss the financial benefit of other companies like Network Solutions.

Network Solutions opened the door to the world that has changed lives forever and created enormous economic wealth today and for the future. The fact that an African American company was the first company to introduce and control the Internet more than 25 years ago is a tremendous accomplishment for our community. Our role cannot be diminished as an integral part of the history of the Internet and its greatness today.

Read the true story of who helped to build the Internet and how society has benefited. Every home, school, and institution should have a copy of this important and historical book to memorialize and demonstrate to future generations of African Americans.

George Fraser is the president and founder of FraserNet, a company that promotes social and economic advancement in the African American community. For more than 30 years, George has been a presenter and speaker to thousands of audiences globally on the value of networking and financial literacy for closing the wealth gap in minority communities.

Introduction: Why I Wrote This Book

Race for the Net is the untold story of when African Americans controlled one of the great technologies of our times, the Internet. It was 1993 when access to the Internet was opened to the global community through a contract awarded to a minority-owned business by the National Science Foundation, a federal agency of the U.S. government. They were assigned the task of creating the platform and infrastructure to allow the general public, for the first time, around the world to have access to the Internet. It was the experience and knowledge of the owners and employees of this African American company that made it possible for people around the world to "travel" on the information superhighway (the Internet). I wrote this book to tell the story of how this amazing technology was initially introduced to the world and who was directly responsible for this global feat.

The advent of new technologies during this phase of the 21st century has been astounding. At no other time in our history have we seen changes taking place at all levels of our life which directly impact so many people. Look at the changes that have taken place in the last 25 years with new means of communication locally and globally, with our ability today to interact with people whom we have never met. The convenience of buying products and services without leaving your home or office, and soon the ability to have your cars on autopilot, driving us from place to place has been developed with the assistance of the internet. These advancements will have a greater impact than previous economic revolutions. The two economic revolutions that precede this one, agricultural and industrial, took twice as long to achieve this magnitude of growth and change in our society. This new economic revolution started with the introduction of computers and mobile phones. At the time, we thought their introduction was tremendous; it enabled us to do so much more with fewer time-consuming tasks and to communicate without being tethered to a physical phone line in our home or workplace. These technologies were amazing and established new industries and economic growth for everyone who had the means to take advantage of these devices. Many young people born after 1985 may not be able to appreciate what I am talking about because most young people today think that these technologies have always existed. Some have never seen a typewriter or fax machine in use, or a rotary telephone, so they can't appreciate the changes that have taken place in such a short period in the last few decades. These types of economic evolutions or changes generally take place in every society and generation. However, not at the magnitude, speed, and potential of our technological society today. The Internet has changed the world forever and has created huge economic opportunities for some.

When televisions became the common entertainment platform in every household, we were able to experience seeing events—some of them live—for the first time in black and white, and then in color. The radio became an option, not a necessity, to know what happened in your city, town, or country and ultimately, the world. You received news commentary by reading the newspaper or magazines and later by watching television news shows. During this period in the '50s, '60s, and even the '70s, we were dependent on others to provide us with information.

The major advancement that has taken place in the last 25 years, creating a new technologically focused society with capabilities we never previously thought possible except in *Flash Gordon* movies and early Dick Tracy comic strips (for the baby boomer section).

All these revolutionary changes today are the result of one major technology platform, the Internet, that was introduced to the global community in 1993.

Internet technology changed the world in the late 20th and early 21st centuries and will continue to evolve. In the early '90s, very few people had heard of the Internet, except federal workers and technologists in educational and research institutions. This technology was developed in the U.S. Department of Defense in the '60s with the assistance of several major universities.

Most people using the Internet today are unaware of the history of how the world acquired this amazing technology. Who was responsible for providing us with access to this revolutionary technology?

In my book, *Race for the Net*, I provide the inside story of how the world gained access to the global Internet, and about the company and people who were directly responsible more than 25 years ago. This great achievement of bringing the Internet to the world in 1993 is an untold story. There have been very few if any, books written highlighting the historical journey of this company's achievement. You will recognize the company's name because it still exists today. But what is not generally known is that the original company was controlled and managed by two African American businessmen. The original founders of the company penned the name Network Solutions in 1979, when they incorporated the company with two white partners, as specialists in computer integration and communication technology located in Herndon, Virginia. I am personally familiar with the history of the company and details of how the Internet was introduced to the world.

In my role as vice president of corporate communications, I was responsible for marketing and promoting the global use of the Internet for the first time on behalf of the company.

The legacy of African American accomplishments as inventors and technologists over the last 200 years is still not appreciated and promoted enough in the global community. Inventions such as the traffic stoplight, refrigerated trucks, IBM's gigahertz chip, etc., among other products and processes developed by African Americans have significantly improved America and world economies, as well as enhanced lives and global health.

The success of the film and book *Hidden Figures,* based on the three African American female mathematicians and their accomplishments at NASA that helped to send John Glenn into space more than 50 years ago, is an excellent illustration of this point. This film has been seen by millions of people globally and has inspired a substantial number of African Americans and other youth to consider STEM programs (science, technology, engineering, and math). The story of the original Network Solutions is another historical achievement. More of these types of stories need to be told.

One of the stories featured in *Race for the Net* is that of the amazing Dr. Lonnie Johnson, an African American man who demonstrated persistence and determination to become one of the leading inventors of recreational products and energy solutions. Many have purchased the Super Soaker water gun or the Nerf gun, but most do not know that an African American invented these products generating more than $3 billion in global sales. He is now working on his next billion-dollar product, new energy solutions for our society. Read about Dr. Johnson's next billion-dollar project in *Race for the Net.*

There are billions of people around the world accessing Internet connections today. This number will grow even more since this reflects only 50 percent of the world's potential users. The Internet economy today exceeds $8 trillion and growing, with new technology like artificial intelligence (AI) and the Internet of Things (IoT).

It is important that African Americans understand that they need to actively participate in the future growth of the Internet. Especially crucial in establishing an early appreciation of how the world will change, particularly creating new job opportunities while eliminating many others.

Becoming more aware that an African American company opened the door for the global use of the Internet will hopefully provide greater motivation not only to become "riders" on the information superhighway but to become developers of new technologies to enhance our economic wealth. Across the country, we are already seeing young African Americans taking advantage of Internet technology creating startup enterprises. However, their ability to scale these businesses to the next level will require capital and the support of our community.

To assist with that aspect, I have also provided information on how individuals, families, and friends can evaluate possible investment opportunities with people whom they know. Not investing in or with each other has been a hindrance to our growth. I highlight several reasons why this happens and what we need to do to increase our economic power.

To that end, I point out several growth areas that I believe will become opportunities for those looking to be in business in the future. It is not always the tech areas that offer growth opportunities; many non-tech areas have great potential for growth. Millennials, there is a section in the book exploring how you can impact economic advancement in your community through your use and understanding of social media and other Internet technologies.

I am sure that you will learn something by reading my book that will hopefully open your eyes and those of your children about becoming an entrepreneur and how to control your destiny.

I hope that you will enjoy reading the book and hearing my interpretation of what REALLY happened with the launching of the Internet around the world.

Part I: The Untold Story of the Internet

"Some of the Greatest Achievements in Life Are Untold or Hidden Stories"

Chapter 1: The Awakening

It was a beautiful spring morning when I arrived at the headquarters of Safeguard Scientific, in Wayne, Pennsylvania. Safeguard is a publicly owned venture fund responsible for funding technology companies from the '70s through the 2000s. During the dot-com boom period, Safeguard financed the early Internet companies led by CEOs barely out of college. There were more than 500 million people globally utilizing Internet technology in 2000; millions of dollars were changing hands daily in support of ideas that centered on the use of the Internet. The Internet had created a global economic marketplace that staggered the imagination. I would call this place my home for two years, working as a senior consultant for Jerry Johnson, Executive Vice President during the dot-com boom.

Still sitting in my car arranging my briefcase and coffee, I received a call from a friend, former international banker Clarence Haynes.

"Are you sitting down Al?" his loud voice resounded.

"Yes, I am still in my car about to head into Safeguard's office. What's up Clarence?"

It was at 10 a.m. "Al, it was announced on the TV's financial networks and published in today's *Wall Street Journal* that SAIC (Science Applications International Corporation) has sold Network Solutions to VeriSign."

After a long pause, he revealed some particulars of the deal that would change my morning and my life. The price tag on this surprising sale was $21 billion—the largest sale of a technology company outside the telecommunications industry to date. Even after hearing Clarence's news, I still could not fathom what he had said. *Did he really say $21 billion?* I turned on my car radio, and almost every station within reach of my dial was announcing Network Solutions had been sold to Verisign for $21 billion by SAIC. In 2000, most of the world knew of Network Solutions as a significant player in the issuance of Internet addresses.

I entered the building at Safeguard as a former vice president at the original Network Solutions; colleagues remarked immediately about the news of the sale and wanted to know if I still had stock in the company. I did not own any stock in Network Solutions at the time of this historical sale nor did any of its four founding partners. Most of the money from this event was made by SAIC and other owners of Network Solutions stock then, through the public offering of the company in 1997 and the secondary offering in 1998. Even then, the secondary public offering valued Network Solutions stock at more than $750 million. It made history as the highest value for an Internet company, at that time.

After getting to my desk, I started to wonder what the four owners thought when they saw and heard the news of the $21 billion sale of their company. I sensed they were questioning their own decision in 1995 to sell the company to SAIC for less than $5 million. I was certain they, like me, were kicking themselves for not holding any of the Network Solutions stock. Even if they had sold 99 percent of the company, their 1 percent would have been worth $210 million on this remarkable day, Friday, March 10, 2000.

Getting any work done that day was nearly impossible. I felt this weight upon me, a force that made me want to surrender to gravity and slump onto the floor like a useless bag of bricks. All I could think about was how the original Network Solutions owners and employees had lost out on this deal. They did not benefit as they should have from their building of a company that at five years from its original sale was now worth more than $21 billion. I started to revisit in my head the years of being on the senior management team as vice president of corporate communications and all of the great things that the company had accomplished when it was controlled and managed by its African American owners, Emmit McHenry and Tyrone Grigsby. Even when this historic announcement was made, only a small number of people knew that the company was once an African American-controlled company.

Selfishly, I wished that the owners had listened to me in 1995. I had begged them not to sell the company and, instead, to take the investment from AT&T or the group of investors that cofounder Tyrone Grigsby had been negotiating with before the sale. *Did they regret their decision to sell and was it too early?*

Even today, there remain so many unanswered questions. The next few chapters will provide you with some information on the company and why the four owners agreed to sell to SAIC, before the Verisign sale. Today, when I buy a URL address from Network Solutions, I wonder, *Could I have done more to keep Network Solutions in the hands of the original owners for even another year?*

Why did the African American community fail to recognize the potential of Network Solutions and provide financial support to keep it in the hands of the African American community? Has this attitude changed in the last 25 years? Is our community supporting one another financially in the technology arena?

Regardless of the outcome, the culmination of events up to and beyond March 10, 2000, the historical facts are undisputed: The original Network Solutions, an African American enterprise, incorporated and established in 1979, was the first company ever to provide (URL) Internet addresses to the world. They opened the door to the public Internet of today.

Chapter 2: My Journey Started in Brooklyn, New York

In the spring of 1964, I traveled to Denver, Colorado to tour the University of Denver's campus as a prospective basketball recruit. Growing up in Brooklyn, New York, I never imagined going that far west to attend college. My high school basketball experience was at Erasmus Hall High School in the Flatbush section of Brooklyn, one of the largest high schools in the country at that time with more than 7,000 students. My graduating class was around 2,000 baby boomer students. Besides being the largest high school in the country in the early '60s, it was the second oldest public high school in the country (established in 1787) with outstanding scholars. Erasmus Hall High School was such an exceptional public school during that time NBC's *Today Show* featured it in 1963 for winning more than half of the Westinghouse Science Awards awarded annually to high schools throughout the country.

The school had many noted graduates like Barbra Streisand; Neil Diamond; Stephanie Mills; Will Downing; Kedar Massenburg; Clive Davis; Eli Wallach; Susan Hayward; Jeff Chandler, basketball legend and founder of the Miami Heat; Billy Cunningham, former owner of the Oakland Raiders; the late Al Davis; Bobby Fisher; Chess Grandmaster, owner and founder of *Black Enterprise* magazine; Earl Graves; and many more distinguished alumnae. Erasmus was predominantly a Jewish institution with a small group of Italian, Polish, and African Americans.

My parents, Luther and Thelma White migrated from a small segregated town of Prentiss, Mississippi in the '30s to settle in the Flatbush section of Brooklyn. My older brother, Martin, and I were the only children from my parents' marriage. Although my parents had received little formal education, they were by no means uneducated. They saw the value in education and wanted my brother and me to receive the best education available. They also required us to attend church on Sundays and any other day that the church was open. Despite the adversity and oppression, they encountered growing up in the segregated South, my parents committed to building a strong economic base for the family. Through their hard work, they acquired real estate in Flatbush, and my mom operated her own daycare center for many years. She was a true believer in owning your own business and supporting other businesses in the community.

Big Shoes to Fill

Not only did my mother support my basketball experience, by attending most of my home games at Erasmus, she also was my biggest backer when I started my own company.

I was named after a successful African American businessman in Flatbush, Mr. Albert Thompson. Along with his wife, Belle, Mr. Thompson lived in the same two-family building as my parents when I was born. Mrs. Thompson drove my mother to the Brooklyn County Hospital for my birth on April 3, 1946. Since the Thompsons did not have children of their own at the time, my mom proposed naming me Albert after Mr. Thompson. Having his name and knowing his business success as an African American businessman in Brooklyn and across the country stimulated my interest in business. My mother had witnessed the success of my namesake, who had built major businesses in Brooklyn in the '50s and '60s. By naming me after him, my mother validated the confidence she had in me to follow Mr. Thompson's footsteps and make a name for myself.

Albert Thompson was from South Carolina. Like my parents, he and his wife came to Brooklyn, seeking change in their lives. He started with a one-truck moving business in Flatbush and grew that business to several trucks, with his name eventually on them. He later learned interior design at night school and opened his own furniture store in Flatbush. Mr. Thompson became very well known in Brooklyn and later throughout the country because he carried a very prestigious line of furniture. In a short time, he even convinced Drexel, a high-quality furniture manufacturer, to allow him to carry their line of furniture even though they had not met him. Before they learned he was African American, Mr. Thompson became one of Drexel's largest distributors in the country during the '50s and '60s. Also, he was renowned for interior design, providing services to famous African Americans during this time. Wanting Al Thompson to design the interior of their homes and churches were people like his close friend Jackie Robinson, the historian John Hope Franklin, John Johnson, founder of *Ebony* and *Jet* magazines, the renowned minister, Rev. Gardner C. Taylor, and many more. Non-African Americans also bought furniture from Mr. Thompson too because he carried high-quality products and provided one-stop design and shipping services.

During this period, the African American community supported each other and kept the dollars flowing within its own communities. I worked at Mr. Thompson's store on Rogers Avenue, and my dad worked as a laborer on Mr. Thompson's trucks—big, bright red trucks with his name, Al Thompson, in big letters. While Mr. Thompson's early business ventures could be operated out of one room, he eventually would own an entire block of offices and a large furniture showroom on Rogers Avenue in Flatbush. The African American community in Flatbush and across the country were very proud of his accomplishments and supported his success. When he moved from Brooklyn to San Diego, California with his family in the '70s for health reasons, we discovered that Mr. Thompson owned one complete block of buildings in Flatbush and had major real estate holdings in Manhattan on West 57th Street. I had some big shoes to fill.

Another business created in Flatbush of considerable reputation was Ebingers, often referred to by many as the best bakery in the country. Established in 1897 by a German family of confectionery geniuses. The Ebingers family chose Flatbush as their place of business and hired people from the Flatbush community. Knowing that Ebingers always had jobs available, my parent's family living in other parts of the city also worked at Ebingers. Most of these family members were also from Prentiss, Mississippi. Eventually, Ebingers in the '50s and 60's operated more than 80 stores in the Metropolitan New York area.

Ebinger's was where my father worked for more than three decades and where my brother and I worked during the summers while in high school and college. When I think about Ebingers, I think about their famous blackout cake as I suspect most people from New York will remember, too. People from all over New York and the entire East Coast would travel to Brooklyn or their other outlets to buy Ebingers' cakes, pies, and many other delicious pastries. If you know a baby boomer who grew up in Brooklyn, mention Ebingers and it may bring tears to their eyes. Ebinger's closed in the '70s, but the memory still makes my mouth water today.

Flatbush was a small, mixed-ethnic community similar to the makeup of the high school I attended. I grew up in the era of Jackie

Robinson, the first African American to enter professional baseball, and it so happened that Flatbush was his home, too. Ebbets Field, where Jackie made history, holds many memories for me, as well. I still remember going to the games with my father as a young boy or with the Flatbush Boys Club to see Jackie play. On any given day at Ebbets Field, seats were filled with many African Americans like my dad and me, attending the game to witness history in the making. My father played baseball for a local team while living in Mississippi. I recall his stories of the other African American baseball players who were members of the Negro Baseball League, of how great they were compared to players in the National Baseball League. On occasion, he would take me to Yankee Stadium in the Bronx to see some of the Negro league teams play, like the Memphis Red Sox, Birmingham Black Barons, Kansas City Monarchs, and Indianapolis Clowns.

Today, it is hard to believe with the large number of outstanding African Americans and other minorities represented in professional baseball and other sports, that it took the courage of Jackie Robinson more than 72 years ago to demonstrate the potential of our community. We were only seeking the opportunity to show our ability to the world and Jackie opened that door.

One of the things that did not exist in Brooklyn during my childhood was overt segregation. Brooklyn was composed of families mostly from Europe, the Caribbean, and resettled families from the American South.

We all worked and played together with no thought of having separate eating places or schools that we could not attend. It was this type of unified environment that played a major role in shaping my life.

Meeting Emmit McHenry at the University of Denver

In 1964, several colleges across the country recruited me for both my basketball and academic abilities. However, academic achievement had not always been part of my story. During my elementary school years, I was a disruptive and distracted student. This resulted in me repeating the fourth grade which hardly ever happens today. This was a wakeup call for me and from that point forward, I was determined never to allow anyone to stop my progress in school, on the basketball court, and later life endeavors. This experience made a big impression even at eight years old; this was a major turning point in my life.

As I pondered which college to attend, my visit to the University of Denver (DU) sold me on the school. I started to look at my life in a different light. A stark contrast from Brooklyn, Denver offered beautiful snowcapped mountains and a refreshing absence of underground subways. I also felt that the University of Denver was the size and environment I needed at that point in my life. While many of my basketball buddies and teammates accepted scholarships to larger schools or schools in the remote West, I felt the University of Denver would provide the educational experience I needed. Little did I know at such an early time in my life, this decision would lead me years later toward one of my life's most influential relationships. At the University of Denver is where I would meet Emmit McHenry.

Emmit was given the responsibility of showing me around the campus during my recruitment trip to convince me to play basketball for DU. The captain of the DU basketball team, the late Frank Mixon, was initially scheduled to provide my tour of the campus, but Emmit, a sophomore student-athlete on a wrestling scholarship, was his replacement. It has been more than 50 years since Emmit gave me a tour of the University of Denver, not knowing how this relationship would change our lives.

My College Experience

Adjusting to college thousands of miles away from home challenged me. My freshman year included a flood of pressure to balance my commitment to basketball while concentrating on my major in chemistry. My early career goals included becoming a pharmacist and eventually owning and operating my own pharmacy. However, the demands of basketball practice and traveling to away games limited the time I could devote to the lab classes. I quickly realized that I could not do both, so I changed my major to a less demanding course of study.

A counselor from the athletic department suggested that I enroll for a career evaluation test, designed to match a person's strengths and weaknesses against specific career areas. I was not surprised to see highlighted as my top career areas related to my answers were as a pharmacist. Knowing this was not going to fit my basketball commitment, I was relieved to see the second area highlighted by the evaluation was business. Therefore, I changed my major from chemistry to business. I majored in marketing and economics at the University of Denver's business school, which was ranked in the '60s as one of the top undergraduate business schools in the country. Since the University of Denver was on the quarter system instead of a semester system, I was able to take more than 60 credit hours each year. When I graduated from DU, I had more than 235 hours of business classes, which helped me gain acceptance to Columbia Graduate School of Business in New York. Choosing Columbia as the next step in my journey meant that I had to sacrifice the opportunity of trying out for the New York Knicks, which had extended me the invitation.

Truth be told, there was a part of me that looked forward to returning to New York and seeing my family and friends. The admissions department at Columbia asked me to visit their office before school started in the fall. I did not know the purpose of the meeting. When I met the admissions department director, she told me that the admissions committee had reviewed my classes and performance at the University of Denver and wanted to know if I would take a course in calculus during the summer. If I successfully completed a calculus class during the summer, I would be exempt from all first-year classes at the Columbia University Graduate School of Business and start as a second-year student.

My faith in a higher being helped me both recognize and appreciate the amazing things that happened throughout my life. The other amazing thing that happened at Columbia was meeting Hughie Mills, dean of the minority student program at the business school. He provided guidance and support for minority students accepted to Columbia Graduate School of Business. In the '60s, minority students attending an Ivy League graduate school was a rare occurrence, but the increased focus on civil rights issues and the Vietnam War made it possible for more African Americans to attend these institutions. The African American class of 1969 at Columbia Business School was one of the biggest at an Ivy League university at the time. Hughie was directly responsible for Columbia's aggressive policy to recruit African Americans. In addition, he was instrumental in the development of a financial assistance program for minority students then and for many years to come.

After Hughie retired from Columbia, I visited him and his lovely wife, Greta, in Las Vegas. During a conversation about Brooklyn, he asked me if I knew of a particularly successful African American businessman named Albert Thompson, who was also from Brooklyn.

Hughie and Thompson had developed a close friendship somewhere along the way and it was at this moment where the three of our lives intersected. As Hughie would explain, Thompson had furnished his house in New York. When I told him that I was not only raised with the Thompson family but also named after Mr. Albert Thompson, a big smile came over Hughie's face.

Hughie often encouraged me to strive to be a standout at Columbia, saying this would impress corporations. He nominated me for the position of student director of Columbia Graduate School of Business Consulting Program, a program sponsored by Arthur D. Little, a major consulting company at the time. The program provided business students with the opportunity to assist small and minority businesses in the Harlem community with services including accounting, marketing, and business strategy assistance. Many of the program's clients would move on to become major business operators in New York. There were more than 15 students who participated in the consulting program and received a small stipend for their service. During my leadership, Columbia's MBA Consultants Program was featured in the *New York Times* as a model for other community business assistance programs.

This accomplishment was the direct result of being mentored at Columbia University by Hughie Mills, someone who had a caring spirit for members of his community.

Emmit and I followed each other's paths after graduating from DU. I was still attending DU when Emmit graduated in 1966. Two of Emmit's classmates and close friends were Earl Brotten and Walter Gibson; these former track stars stayed after graduation to take graduate-level classes. They were my roommates and kept me informed of Emmit's activities.

Emmit enlisted in the Marines and then enrolled in graduate school at Northwestern University in Evanston, Illinois after suffering a back injury in the service. He had an illustrious career as an insurance executive with major Fortune 500 companies while I became a banker with JPMorgan. JPMorgan recruited me prior to my graduating from Columbia University since they knew in my last year that I only needed to take independent study classes in order to graduate. Since these courses did not take up most of my time, they made me a job offer to start several months before graduation. At the time, I was uncertain as to what area I wanted to pursue in banking, but I knew from my education at the University of Denver and Columbia, that I wanted to learn more about money and how to make it. To this end, I believed the best path forward included working in a bank and JPMorgan's offer seemed to fit the bill. My starting annual salary in 1971 was $14,000 (at that time, a lot of money!).

JPMorgan Experience

I had heard of JPMorgan while attending Columbia but did not know the history of its original founder as being one of the wealthiest people in America, with the capability of financing the U.S. government and several major industrialists during World War I. From the moment my feet hit the marbled bank entrance at 23 Wall Street, I was over-come by the regal and stoic setting of bankers sitting at their desks in three-piece suits.

At the time, the banking platform consisted primarily of white men, mostly graduates of Ivy League schools, with only a few white women and even fewer African American men or women. I was not aware of the lack of diversity when I accepted my first assignment in the international division. I was only the second African American the bank had ever hired to work in this area. The other African American who preceded me in the international division was Ralph Bunche Jr., whose father was the U.S. delegate to the United Nations. Ralph Bunche Sr. was responsible for negotiating the Middle Eastern peace agreement.

I did not choose the international division; rather, it chose me. JPMorgan's human resource department reviewed my background and saw that I had served as director of Columbia's Business School Consulting Program and that I had been featured in the *New York Times* during that tenure. I had also been a community business consultant in Denver while at the University of Denver. For these reasons, JPMorgan asked me to accept a position in its International Trade, Commodity, and Ship Financing Group. Concerned about my lack of previous experience in the international area and feeling more inclined toward a commercial lending officer position working with major Fortune 500 companies, I reluctantly accepted the position following my completion of the bank's training program. This is one of the best decisions that I did not make for myself. I was very young, around 25 and had no experience working in a major financial institution. What I did know was that my classmates at Columbia had assured me that my decision to work for JPMorgan would impact my career for the rest of my life and that they wished for my opportunity. During my time at JPMorgan, I met the late Myron Taylor, an African American who was considered a genius at understanding financial markets as well as trading U.S. Treasury securities. Myron Taylor, a graduate of Howard University, was mentored by the head of the U.S. Treasury Division who taught him all the skills he needed to run a multibillion-dollar trading desk. I was very impressed with how JPMorgan provided this fellow African American the opportunity to become a major player in the bank and ended up being someone who influenced me during my time at the bank. Myron departed the bank when he failed to receive just compensation for a transaction; a multimillion-dollar deal for the bank and for which he only received a check for $25,000, far below his expectation and lower than the compensation received by other managers. He later went on to become a financial advisor to one of the oil-rich countries in the Middle East. Myron's experience at JP Morgan made it possible for him to attain this high-income position later in his career.

After the training program, I was introduced to George Cashman, VP, and head of JPMorgan's International Trade Finance Group, with whom I would be working directly at the bank. He taught me how to finance multimillion-dollar power plants, aircraft, mining equipment, tractors, and various other products and services utilizing the financing program of the Export-Import Bank of the United States and other global trade financing facilities. He would take me out on calls and introduce me to treasurers of some of the largest corporations in the country. I was a sponge. I listened and learned from him how the bank could finance a $100 million transaction in Poland or a $350 million telecommunications project in Spain. After one year, I had developed my own accounts with Fortune 50 companies' clients in the Midwest that manufactured drilling equipment and earthmovers. Some of my clients included Caterpillar, International Harvester, John Deere, General Motors, and many others.

After approximately five years at JPMorgan, I started to feel a little antsy, since I had not been promoted to assistant treasurer, which was the first management-level title. Furthermore, I felt that if I were to make a career in international banking, I needed to gain some international lending experience. This entailed being able to make credit decisions on country lending lines to foreign governments and commercial customers abroad. The bank wanted me to continue with my work in their international trade group, but this was a staff position and not a line position where I could be promoted in less than 10 years. This was the traditional waiting time in the '70s before JPMorgan would offer management titles. My request to be moved to a country lending position was handily denied. The company expressed its appreciation for the great job I was doing in the international trade group, as was evidenced by high customer marks on deals I structured. Nevertheless, I could not shake the feeling that they pigeonholed me in a position where I could not expand my international capabilities. Also, I still remembered how the bank treated Myron Taylor, but he was able to leverage his Morgan experience elsewhere.

By this time in the history of the banking industry (the mid-'70s), other banks were willing to pay a premium for trained African American bankers. I was receiving calls from recruiters every day asking me if I was happy and offering me positions in international banking at other institutions, that would provide a new platform of personal growth, an officer title, and country lending authority. At the age of 29, with a wife and two children, I resigned from JPMorgan.

I remember the JPMorgan HR person saying, "Al you can have a great future at JPMorgan, but you need to wait your time." I believed my time had come, so I received and accepted a substantial offer from Bankers Trust Company (now Deutsche Bank).

I was appointed as the assistant treasurer in the Asian Division, lending to the Philippine government and the largest commercial Philippine company, the Soriano/San Miguel Group. My Filipino clients would often ask me about my basketball background since they had a love of basketball. I would mention the time I played against Kareem Abdul Jabbar (then Lew Alcindor) in high school when he was a freshman at Power Memorial in New York. They wanted to know if I blocked his shot, then we would all laugh together.

My time at Bankers Trust Company provided me with great experiences working with foreign businesses to help fund their projects in their countries as well as throughout the world. The perks were not too shabby either, at times including a case of San Miguel Beer and cigars each Christmas. Before I could make my first trip to the Philippines, Bankers Trust management asked me to consider a new assignment, creating the International Finance Group within the bank. Due to my extensive experience at JPMorgan and my reputation in the industry, they wanted me to consider this new endeavor. It came with another title, more money, and my own staff reporting to me. This leadership role was not new to me since I was the captain on every basketball team I played on during my academic life.

My expanded role at the bank also included working with foreign branches of Bankers Trust Company in Europe and Asia. My group had become recognized throughout the bank, and other international banking divisions began pursuing my team members for their group. I had trained my staff in how to structure and finance international trade transactions utilizing the programs of the U.S. Export Bank (Eximbank) and trade finance programs in Canada, England, France, Italy, and Japan. These programs were established to finance sales of U.S. and foreign-made products of their respective countries.

During my tenure as head of international finance, I also wrote manuals on how to structure various types of trade transactions. This advice and financing assistance I provided to my international clients were making them millions. I became very envious of my clients because of the wealth I was creating for them. While perks like swanky lunches at Park Avenue restaurants and meeting famous people continued to sweeten moments occasionally, deep inside I wanted more. My clients liked to introduce me to their friends, saying, "This is Al White, my banker." I wanted to have my own banker and be able to pay for my own lunch at exclusive restaurants of my choosing, so you know what I did? I got going after four years.

Leaving Banking for New Opportunities

Although my friend Emmit continued to work in the insurance industry, we regularly communicated about the possibility of him starting his own company. In 1979, that dream was realized when he and his partners formed Network Solutions. He did not participate in the initial daily management until 1987 when he left the insurance industry.

After my 10-year banking career, I formed my own international trading and consulting company and soon found myself serving as a financial advisor to the late Percy Sutton, owner of the largest network of African American radio stations in the country. I worked with the late Reginald Lewis and Wally Ford, who is today a prominent attorney and college lecturer in New York who also supported Percy Sutton's empire. One of my other clients and a partner of Mr. Sutton was his attorney, the late John Edmonds. He owned several mining properties in Mexico and asked me to review his operation plan for advice on how to acquire equipment for his Mexican silver mine. One of John Edmonds' relatives was someone who would become one of my closest friends. Joseph Searles was a former New York Giant football player, who, after his retirement, became a partner with an investment company in New York. Joe was appointed to a seat on the New York Stock Exchange, the first African American to garner this position. Joe would tell me stories of how Wall Street executives had a difficult time accepting him into their exclusive circle although being a former professional athlete helped him in some ways.

I had other clients in the international arena as well as contracts with federal government agencies including the U.S. Agency for International Development (USAID), the U.S. Department of Commerce, and the city of Miami. During this time, I continued to sharpen my entrepreneurial and finance skills.

During my tenure working for Percy Sutton, he requested that I attend an advertising shoot in Harlem for the radio stations. When we arrived at the commercial shoot, there was a new black Bentley shining from the sun and the famous model Beverly Johnson sitting in a chair waiting for Mr. Sutton. Cameras surrounded the car and Ms. Johnson. Inner City Broadcasting was trying to present a new image for the radio stations around the country. Mr. Sutton wanted to know if I had any ideas for the shoot. I suggested that he have Ms. Johnson slink her body across the hood of the Bentley with her mink coat over her shoulder. They shot several takes from my suggestions and this ad was seen on billboards and buses throughout the country with the title "Touch of Class." This became one of Inner-City Broadcasting's most successful promotional campaigns during my tenure with Mr. Sutton.

In 1985, I moved to Washington, DC to be closer to some of my federal clients. Emmit mentioned that he was leaving his insurance position and eventually moved to Washington to assume the CEO position at Network Solutions. This proximity between two old friends would provide an opportunity for us to see each other on a regular basis and share experiences from our individual companies.

Chapter 3: The Power of the Internet

Major Technology Contribution to Building the Internet

As with every great technology breakthrough, there are always a few individuals dedicated to going beyond the typical contributions in order to make a difference. The development of the Internet is no different.

The computer networking revolution began in the early 1960s and has led us to today's technology. The Internet was first invented for military purposes, and then expanded to the purpose of communication among scientists. The invention also came about in part by the increasing need for computers in the 1960s. During the Cold War, it was essential to have communications links between military and university computers that would not be disrupted by bombs or enemy spies. In order to solve the problem, in 1968 DARPA (Defense Advanced Research Projects Agency) made contracts with BBN (Bolt, Beranek and Newman) to create ARPANET (Advanced Research Projects Agency Network). States in article by Alaa Gharbaw UCSB 1991 (Revolution of the Internet)

The first breakthrough was 50 years ago, on October 29, 1969. Bill Duvall, a scientist at Stanford Research Institute (SRI) and a graduate student, Charley Kline at UCLA wanted to send a message over the ARPANET. They were more than 350 miles from each other in California. Even though their computers crashed, they were able to send two letters "L "and "O." These two letters provided proof that it was possible and feasible to send a message to dissimilar computers. At the time, they thought nothing much about this achievement, but this was many believe the beginning of the Internet. "It allowed the distribution of virtually all of the world's information to anybody with a computer," states Mark Sullivan in his industry article. Following this achievement, in 1971 the first e-mail was sent by an MIT researcher, Ray Tomlinson.

During the next decade, a brilliant British scientist, Tim Berners-Lee, developed the World Wide Web, a platform that allowed those on the Internet to access information from other sites. When the general public gained access to the Internet this technology became invaluable for the billions of people seeking access to information. Another major contribution was made by two engineers also working with the Defense Department, Vinton Cerf, and Robert Kahn; these individuals are considered two of the "fathers of the Internet." Their technology solution allowed computer networks around the world to talk to each other with their development, communication architect of the Internet. Transmission Control Protocol, or TCP. (Later, they added an additional protocol, known as Internet Protocol.) One writer describes Cerf's and partner's protocol as "the 'handshake' that introduces distant and different computers to each other in a virtual space.

I have provided you a brief summary of the building blocks of the Internet and contribution by some great people over the years

If more technical details on how the Internet was built and designed are needed, I suggest that you read the book published by the Foundation for Enterprise Development entitled, *Names, Numbers and Network Solutions, the Monetizing of the Internet.* It provides more detail of the Internet's development.

People knowing that I was writing a book on the history of the Internet, always asked the question, "Did Al Gore invent the Internet?" He did not invent the Internet, as you have read, but played a critical role in the commercialization of the technology in 1992.

However, in the '90s another major contribution occurred that will be discussed later in the book, assisted in the proliferation of the Internet we know today.

Today's Internet

Very few technologies have had the impact of the Internet on our world. It has revolutionized how we communicate with individuals, companies, and numerous other entities. It is considered today the most powerful tool in the world. It has made access to information and communications far easier. Many believe that the Internet is the ultimate disruptive technology of our time.

Disruptive technology is one that displaces an established technology and shakes up the industry or a ground-breaking product that creates a completely new industry.

Many young people today don't understand how we survived before the Internet. For those who are old enough, you remember going to a local library and going through card files to identify a book or enough reference material to write an article or school paper. This was a time-consuming event that made these tasks much less attractive to do unless there was someone at the library you wanted to meet! Today, you can access this information from computers, smartphones, or tablets anytime and anywhere. What about meeting a new friend who could possibly become your future husband or wife? Back in the day, you waited for your friends or relatives to introduce you to potentially one or two suitors. In many cases, you would eventually have to arrange a meeting through a phone conversation and without knowing what the other person really looked like. In the old days, this was called a "blind date".

Wow, that has changed dramatically in the last 20 years! Just sign up for one of the hundreds of dating sites and be matched with 10 or more people who match your profile. Get a picture instantly of the person and profile of their background and even their credit rating. The Internet is the primary meeting space for couples, either casual or more than casual: Today, 70 percent of relationships start online, states Reuben Thomas, a sociologist at the University of New Mexico. "The majority of newlyweds today met on the Internet and is growing," said writers at *USA Today*. As mentioned, a high percentage of Internet users have established a relationship and a date through this technology platform. There is no age limit in using matchmaking services; this is a service that caters to everyone from young people to senior citizens. If you have an Internet connection, you can find someone. This business is generating more than $4.6 billion globally.

Even Facebook is now entering this lucrative market with its own online matchmaking services. Generally online sites provide introductions to strangers. Facebook under its structure of billions of users has a program to assist you to link to friends of friends. Remember Facebook has a profile on you and everyone in your family and their friends. They are going to utilize this information to identify a partner in your database of close associates. Many experts believe that Facebook's service could be huge since they already have your profile. Yes, you will have to give them permission to use your information for this purpose.

One of the more interesting reports in 2013 by researchers at Harvard University and the University of Chicago showed that marriages that started online were less likely to end in divorce and were associated with higher levels of satisfaction than marriages of couples that met offline. These results may be due to knowing more about the person before meeting them (credit, employment history, prior relationship information, and other details). This is information that the Internet can provide before establishing a serious relationship. With the advent of Artificial Intelligence (AI) technology, finding the right match has improved substantially.

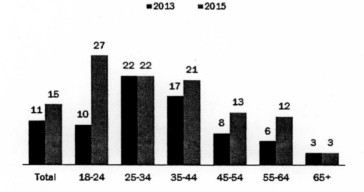

Use of online dating sites or mobile apps by young adults has nearly tripled since 2013

% in each age group who have ever used an online dating site and/or mobile dating app

■2013   ■2015

Source: Survey conducted June 10-July 12, 2015.
PEW RESEARCH CENTER

This chart demonstrates the growth of dating sites or mobile apps by age group from 2013 to 2015.

Remember going to the mall or a retail establishment to buy something to wear to that party or an event? Getting in your car or walking to a store to find something to give a friend for their birthday or anniversary? These were time-consuming tasks. Today you can get those items on the Internet as well as purchase hundreds of other items and services and all can be delivered to you in one or two days. What about finding a doctor or dentist who is located near your house or getting directions to your next meal at a restaurant?

Did you know that a large percentage of the population relies on the Internet for medical advice and pharmaceutical information? Most people utilizing the Internet for this type of information are seeking guidance and opinions on how to treat minor illnesses or wanting additional information on treatment or drugs prescribed by their doctor. I am sure you have researched a medical treatment that you thought too minor to ask your doctor and be billed for. Before the Internet, most people would ask an older relative or friend if they had any ideas on how to treat a minor ailment, like a sprain of your leg or back, stopping chronic hiccups, or a cough. The biggest investigation on the Internet is cross-checking prescription drugs that your doctor may have ordered for you. Many medical experts do not advise using the information on the Internet to treat life-threating illnesses nor accepting all the advice you read on medical treatment on the Internet "as gospel." I personally agree, you are not a trained doctor and you do not know if the online information is suited for your specific medical problem. Most studies have shown that baby boomers utilize the Internet far more for medical advice than younger people. This may be directly related to the cost of a visit to the doctor or the medicine they are taking is too expensive; they are seeking other options to solve their medical issues.

Before the advent of the Internet to find a friend from high school or ex-girlfriends or ex-boyfriends it was almost impossible unless you followed their address or telephone over the years. If you were looking for a lost sister or brother, or even your natural mother or father because of being separated due to an adoption, it would take many years of research and discussions with relatives to identify where you could find these family members, if at all. Now, through the technology of the Internet and genetics, your chances of locating someone from your past have increased.

While writing this book, I had a personal experience with the ability of the Internet to unite people. My older brother, Martin, as a young man living in Brooklyn, fathered a baby boy. He never had the chance to see the child since the mother put him up for adoption. On many occasions, he would mention the son he had never met or seen and wondered how he had grown up. On April 25, 2018, I received an email from a man named Mark Jackson who asked me a couple of questions about my family. He thought that he may be related to me. He had submitted his DNA on www.ancestry.com and it matched my brother's DNA. Then Mark had to find the location of Martin L. White. He searched Facebook and found several Martin Whites, but only one ever lived in Brooklyn, New York. With this information, he saw my name and profile on Facebook, as being Martin's brother. Following a brief discussion, I realized he was my brother's biological son, whom he had never seen in 57 years. One of the reasons he contacted me was due to my physical location, in Silver Spring, Maryland where he also lives with his family. In fact, we live less than 10 minutes from each other. This was a wonderful and joyous reunion for my brother and our family provided by technology that I was associated with from the very beginning, the Internet. We were able to share Facebook pictures of family members and Mark could make contact immediately with his sisters and cousins he had never met.

Thousands of people have used this combination of amazing technology to bring people together with loved ones anywhere in the world, all brought to you by the power of the Internet.

Facebook reaches more than 1.9 billion people daily. If you include the other social media sites and other search platforms, the chance of locating a person you are looking for and their last address and telephone number is made possible from the Internet. Yes, on the other side of the coin, it has made it easier for creditors to find the location of delinquent borrowers.

You get my point; the Internet has become one of the most powerful tools on the planet today.

In the final days of one of the worst disasters in the history of our country, Hurricane Harvey, that paralyzed most of Houston and the surrounding areas dumped trillion tons of water, 100,000 houses were destroyed. This also left thousands of people homeless. Unlike Hurricane Katrina more than 1,800 deaths occurred because of the lack of communications in 2005, Hurricane Harvey had less than 100 deaths. However, even one death is a tragedy. This technology helped save many lives during Hurricane Harvey. Many people stuck in their homes and businesses used Facebook and Google Maps and numerous other tools/platforms that can be accessed by the Internet to provide a location or rescue information or distress call. Even the fundraising process following this disaster was made easier by access to the Internet. The head of FEMA (Federal Emergency Management Agency) asked that people with Internet access make donations to various disaster relief websites: www.nvoad.com and disasterassistance. org. The Internet's global communications network has assisted in the aftermath of this disaster to raise hundreds of millions of dollars through donations and contributions, but more importantly, it saved thousands of lives.

One of the other benefits of the Internet is its ability to teach new skills without sitting in a classroom. This may be the most profound use of the Internet in the future since we will need to train and educate a new segment of our population who will be negatively impacted by the loss of employment with the advanced capabilities of the Internet. Yes, like most economic revolutions in our country, we have achieved substantial productivity improvements or time-saving gains at the charge of lost jobs. There are negative impacts of the Internet, but I am sure there are few people today who would replace the benefits received by the Internet over the last 25 years. The Internet has also created many technology companies that have added products or services that have changed the way we live our everyday lives and continue to effect change. Many experts believe we are just at the beginning of the next Internet revolution. The next phase of the Internet's development will develop to generate more than $25 trillion in economic growth with more devices and labor-saving technology.

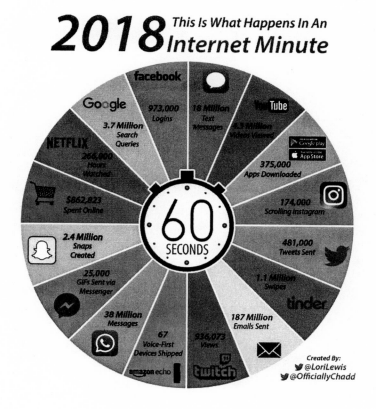

At the advent of the public Internet in 1993, there were few who knew or understood the potential of the Internet and what profound impact it would have on the global society of the future.

Then there were others who clearly saw the potential and started to establish companies that could take advantage of this universal communications platform. Many of these companies are successful billion-dollar empires today. In 1994, Jeff Bezos, the founder of Amazon, was one of those people who was an early adopter and saw the potential of the Internet and its global communications reach. Bezos used the Internet to create the Amazon platform to initially distribute books, and now so many other products and services.

This powerful tool is based on an addressing system that has today more than four billion addresses of individuals, companies, organizations, and institutions providing communications everywhere in the world.

Even emerging countries around the world are benefiting from the technology of the Internet. In his article "How the Internet Is Changing Life for the World's Poor People,"

Christopher Mims states that entrepreneurs are devising new services to provide neighborhood-scaled renewable energy and clean water, gas-cooking stoves, microloans for consumer goods, and insurance against natural disasters.

Additionally, in countries with emerging economies, the Internet is providing social media, video, and communications from the outside world. In the past, many inhabitants in developing countries had limited access to movies and video except for the family TV, that the entire family watched. With the Internet and cell/smartphones, family members have more individual freedom to watch shows they want to see on YouTube, podcasts, as well as streaming systems. This type of freedom has opened the door for poor people all over the world. Most people believe the greatest future growth will be in emerging markets where the Internet is exposing common people to things, they never thought possible for them to experience. As mentioned, learning new things will provide enormous opportunities for children and families in developing countries to educate themselves, without having to travel for hundreds of miles.

One of the major benefits of the Internet for me, and I am sure, for millions around the world is the streaming of religious services, especially sermons of some of the leading preachers and ministers anywhere in the world. In many cases, you can see the live services on your computer, smartphone, or tablet. If you miss the service, especially the sermon or speech, just go to YouTube and enter the person's name and most of their presentations can be found and watched. In the past, you would have to get a copy of a CD or DVD to replay a recording of the event.

Also, today, you can contribute to your religious denomination online when you are unable to attend the service. This is made possible by the Internet.

In fact, I recently went to Best Buy near me to purchase a CD I wanted to hear and was told by the salesperson that they discontinued selling CDs since everything is being streamed on the Internet and can be downloaded on your smart device. Best Buy does not want to make the mistake of the Blockbuster store several years ago when they did not respond quickly enough to advice of market experts when they stated that there would be no need to rent movies since the Internet would enable people to download movies in their homes.

Finally, the Internet's greatest contribution is the ability of every human on earth to be able to freely express themselves. No longer must you be a trained newspaper or magazine reporter, on a TV or radio news commentator to express your viewpoint to the public. The Internet allows everyone to be able to provide their opinions or thoughts to the world. This technology provides an open communications platform making the first amendment of the United States Constitution, (freedom of speech), a reality for every American. The Internet allows people—children, adults and millennials, regardless of race, sex, or religious origin—to openly communicate to anyone who has access. Even major political figures utilize its various platforms to present their points either negative or positive. You can send messages to political adversaries, regardless of the time of day and they will know instantly what you are thinking or planning to do. A political leader's strategies and opinions can be communicated on Twitter (up to 280 characters) while you are sleeping.

Chapter 4: The Founding of Network Solutions

In the '70s, Tyrone Grigsby and Emmit McHenry met at a meeting of minority insurance executives of the Ujamaa organization and they shared a common interest in the advancement of their careers. Tyrone Grigsby (Ty) was a graduate of Howard University and grew up in the Washington, DC area. Ty was the eldest of eight brothers and sisters. He served in the U.S. Army Reserves and held various jobs in DC with several federal agencies.

After graduating, Ty accepted a position as an investment analyst in the private placement group at Travelers Insurance in Hartford, Connecticut. While in Connecticut and on leave from the company, he became the executive director of the Ebony Businessmen's League in 1971.

He stayed in the New England area for a couple of years before returning to Washington, DC in 1973 where he took a position with a computer firm, Raven Systems and Research. As executive vice president and a shareholder, he was able to grow Raven System's annual sales from $250,000 to $10 million. He left the company after five years. Ty acquired an understanding of the government contracting business and developed a relationship with a major DC law firm that later assisted him with meeting key people at federal agencies and applying for the Small Business Administration (SBA) 8(a) minority business certifications needed to get preference on federal contracts for his future employer.

He had been in touch with Emmit McHenry after leaving Hartford and the two discussed starting a company of their own.

Both agreed that it was a good idea, and Ty kept in the back of his mind the success of Raven Systems in the federal space as a minority-owned company. He started to investigate opportunities and met with Sam Harrison, a headhunter, who mentioned that he knew two individuals, Gary Desler and Ed Peters who were also interested in starting a similar company.

They arranged a meeting, and they found that the team fit well because they all had an interest in the same areas of technology. They discussed their computer backgrounds and experiences. Emmit mentioned that he had been a system engineer with IBM and had been trained on mainframe computers. Gary and Ed had hands-on technical experience and relationships with the U.S. Department of Labor where they thought opportunities existed for their new company.

Gary Desler grew up in Omaha, Nebraska and was the eldest of three children. At an early age, he was stricken with polio but survived to become a competitive athlete in three sports at his high school. His early business experience was working on his aunt and uncle's farm for four summers and selling their fresh eggs door-to-door. He also worked five summers on a construction site overseeing projects. After graduating from high school, Gary joined the U.S. Navy where he began his technology training in computers and radar systems equipment. After the Navy, he was hired by Digital Equipment Company to train on timeshare computers and systems engineering.

He worked at the Central Intelligence Agency (CIA), National Security Agency (NSA), and many other federal agencies while at

Digital Equipment Company. He enrolled in Prince George's County Community College in Maryland and later at the University of Maryland where he obtained a degree in information technology and business. He worked at several federal agencies after graduation and became an expert in mainframe computing and systems. He later accepted a position at the Department of Labor in their computer and IT division.

Ed Peters grew up in Ft. Lauderdale, Florida after his mother moved from the eastern shore of Maryland following the death of his father. He was the oldest of three children. He attended elementary school in Florida and following graduation from high school, was accepted at Catholic University in Washington, DC. After graduating, he worked for several technology companies but finally settled on a position at the Department of Labor as a systems integrator.

This diverse group of owners worked together for more than 16 years and built an impressive technology company, but it was not easy.

The initial name of the company, Network Solutions, was originally coined by Gary Desler, one of the white partners, for another company that he'd planned to operate. However, after meeting with Emmit, Ty, and Ed they decided to use the name Network Solutions, Inc. (NSI) that Gary had reserved.

It was January 1979 when two African Americans, Tyrone Grigsby and Emmit McHenry, the majority stock owners, formed Network Solutions with Gary Desler and Edward Peters, two white, minority stock-owning partners. The company was formed to focus on data engineering and integration. This was an area in which all the partners had a relevant background and understanding.

After forming the company, the four partners decided that they should apply for certification in the SBA 8(a) program. Since the company was majority-owned by the African American partners, they would be eligible to participate in this program and receive special treatment under this certification.

Background on SBA 8(a) Program

In 1970, the first of these regulations articulated the SBA's policy of using section 8(a) to "assist small concerns owned by disadvantaged persons to become self-sufficient, viable businesses capable of competing effectively in the marketplace." A later regulation, promulgated in 1973, defined disadvantaged persons as including, but not limited to, "African Americans, Spanish Americans, Asian Americans, Eskimos, and Aleuts." However, the SBA lacked explicit statutory authority for focusing its 8(a) programs on minority-owned businesses until 1978, although courts generally rejected challenges alleging that the SBA's implementation of the program was unauthorized because it was "not specifically mentioned in the statute." The 1978 amendments also defined economically disadvantaged individuals, for purposes of the 8(a) program, as "those socially disadvantaged individuals whose ability to compete in the free enterprise system has been impaired due to diminished capital and credit opportunities as compared to others in the same business area who are not socially disadvantaged."

In addition, 8(a) firms must be controlled by one or more disadvantaged individuals. "Control is not the same as ownership" and includes both strategic policy setting and day-to-day management and administration of business operations. Management and daily business operations must be conducted by one or more disadvantaged individuals unless the 8(a) business is owned by an Alaska Native Corporations (ANC), Certified Development Company (CDC), Native Hawaiian Organizations (NHO), or Indian tribe. These individuals must have managerial experience "of the extent and complexity needed to run the concern" and generally must devote themselves full time to the business "during the normal working hours of firms in the same or similar line of business." A disadvantaged individual must hold the highest officer position within the business. Non-disadvantaged individuals may otherwise be involved in the management of an 8(a) business or stockholders, partners, limited liability members, officers, or directors of an 8(a) business. However, non-disadvantaged individuals may not exercise actual control or have the power to control the firm or its disadvantaged owner(s) or receive compensation greater than that of the highest-paid officer (usually the chief executive officer or president) without the SBA's approval.

Network Solutions was able to meet all these tests for approval with Emmit McHenry and Tyrone Grigsby being the majority shareholders and having management control of all decisions.

Tyrone was elected president of the newly formed company and Emmit, the chairman of the board. While three of the partners were working full time in the company, Emmit, for personal reasons, decided not to initially join the management team of the company, but instead held the position as chairman of the board. Even though he was not participating in the day-to-day running of the company, he participated in most business decisions related to the operation of the company. As most new business owners know, starting a business from the ground up takes determination, capital, and luck. Initial capital for a startup comes from the founders, friends, and family, as well as personal credit cards. Network Solutions' partners faced all these issues and more to keep the company afloat in the early days. Through Ty's relationship with a local banker, he was able to borrow $10,000 to assist with the startup. Even with the loan and family/friends' financial support, the company struggled in the early days. During this period, the owners drew no salary while trying to grow the business.

Going into 1980, they won a contract for $300,000 and hired eleven employees. The company continued to struggle in 1980 and showed a net worth of negative $90,000. They were unable to pay their employee taxes, resulting in some doubt concerning their survival. However, Gary Desler through his dentist contact was able to get a bank loan to assist in paying down the taxes and continuing in business. They received a credit line from the bank for $150,000 which later grew to $6 million over the next five years.

Eventually, Tyrone Grigsby, Gary Desler, and Ed Peters won an opportunity with the Department of Labor that put the company on track for garnering more opportunities. It is very difficult for new companies to win contracting opportunities without showing past performance or owners having a background in an area of service. Most federal agencies and even commercial customers do not want to take a chance with new companies and a management team that lacks prior experience in the area of need.

From the Small Business Administration's statistic of new companies' failure, most startup firms fail within the first two to three years of operation. From my own experience, startups face enormous challenges from all sorts of issues, but the major one is the lack of capital.

Besides capital resources, timing is also critical. If a business is unable to take advantage of their window of opportunity within some reasonable time, it can face major problems of execution. Most small businesses have unrealistic goals when they start, thinking that business "will just come in the door when they hang up their shingles."

However, customers and suppliers do not want to initially provide support to a new company because they want to see if it can survive beyond several months. Additionally, financial institutions are reluctant to lend to businesses with no previous track record or assets. Especially in a service business like Network Solutions; they had no assets to pledge except their personal homes or land. Financial institutions (if you have the assets) want personal guarantees of the founders and perhaps even their spouses. Network Solutions' founders in the early days (and even later) had to provide those guarantees to raise funds.

The founders of Network Solutions received a small contract of $1 million in 1982 and ended the year with sales of $1,100,000. They continued to build their business base under their 8(a) certifications and would receive other opportunities from federal agencies. However, they were still not profitable until the third year of operation. Network Solutions continued to gain some traction with the technical background of the partners and relationships with directors at several federal agencies. They were able to hire a talented team of managers with capabilities in some of the new technologies in the software and hardware areas.

One of the major problems of computerization back then was getting mainframe computers to "talk" to desktops. This impeded many federal and commercial organizations from taking advantage of the storage power of mainframes since there was no communications link. Network Solutions' owners and management team had developed software that could solve this problem and were able to team with IBM to win a major contract at NASA's Johnson Space Center in Houston, Texas to demonstrate their capabilities. John Marshall, one of the lead marketing managers at Network Solutions, stated that "winning this contract with IBM at NASA and unseating Computer Science Corporation (CSC) was critical to Network Solutions' growth in the '80s." Since federal agencies were supporting a mixed-team approach of large primes (prime contractors) teaming with small businesses, IBM was able to provide a winning proposal that NASA accepted over the incumbent.

By 1985, six years after their formation, the company had reached a revenue level of $18 million with a substantial backlog of business. With the IBM contract and some smaller ones, they were building a past performance record that helped them to receive other business. Without good past performance credentials, it is highly unlikely anyone will use a service or product. The exception is when you have a service or technology that is above the performance expectation of what the client would gain from other competitors. In that rare circumstance, they will take the chance; but not for long. The product or service must save them money and time. Most companies will provide a pilot or simulation of their solution, so the potential customer can evaluate the technology before they buy.

Network Solutions was able to stay ahead of their competition with these types of solutions for system integration to software development. The company was always able to demonstrate their superior value to their customers. Once they established a superior reputation in the federal sector, they found other opportunities in the commercial space. They landed contracts with Bank of America, Nations Bank, the state of California, and many others.

In 1987, the management of Network Solutions changed hands with Emmit McHenry leaving his executive position in the insurance industry to become president and CEO of the company, replacing Tyrone Grigsby who went on a sabbatical. In most companies, this type of transition is normal. The person that leads the company and develops the initial strategy and direction often becomes burned out trying to keep the company running.

Emmit brought with him fresh ideas and relationships in the corporate world that Tyrone did not possess. Tyrone had done a great job of keeping the company afloat and helping to develop business in the federal space with the assistance of Gary Desler and Ed Peters. Since the company was growing in revenue and employees, Emmit's background was more suitable to take the company forward.

Emmit McHenry, the oldest of nine children, was born in Forrest City, Arkansas, and was the grandson of a minister. He grew up in Tulsa, Oklahoma where he showed entrepreneurial skills at an early age. He operated several businesses from adolescence, from cutting grass to delivering newspapers. By the time he entered high school, Emmit had become a conglomerate, having his younger brothers and sister assist him with his businesses, as well as other local kids. While working at the local country club, he also came to understand the difference between "old money" and "new money."

Emmit had a cousin in Kansas City who was his role model. The cousin owned several businesses from janitorial services to trash collection, and scrap metal to pig farming. His cousin was focused and owning several companies and his success stayed in Emmit's mind for many years and was the basis of his desire to own a business. While growing up, Emmit was always mindful of Tulsa's dark history of the Greenwood riots in 1921. Greenwood was considered the "Black Wall Street." It was one of the wealthiest and most desirable places for African Americans to live in America in 1921. It was home to dozens of prominent African American businessmen. Greenwood boasted a variety of thriving businesses that were very successful until the events known as the 1921 Tulsa Race Riot. In fact, the district was so successful that a dollar would stay within the district an estimated 19 months before being spent elsewhere. Not only did African Americans want to contribute to the success of their own shops, but racial segregation laws prevented them from shopping anywhere other than Greenwood. Following the race riots and fires that destroyed Black Wall Street, the area was rebuilt and thrived (with more than 100 African American businesses in place) until the 1960s when desegregation allowed African Americans to shop in areas from which they were previously restricted.

Emmit won a scholarship to the University of Denver, where he and I met. After graduating, he was drafted into the military during the Vietnam War. He was in the Marines for two and a half years until he injured his back and left with a medical discharge. He continued his education, attending Northwestern University where he earned his master's degree in communications. He completed the qualifying examination for a doctorate degree.

Emmit's management experience started with IBM as a system engineer and later with Connecticut General and Union Mutual. He ended his insurance career with Allstate Insurance. He distinguished himself in the industry by serving on several industry committees and was a founding member of the American Productivity Management Association.

Left to right: Ed Peters, Emmit McHenry, Gary Desler, and Tyrone Grigsby

Chapter 5: The Greatness of Network Solutions

In articles and a book concerning the original Network Solutions, very little has been written about the enormous accomplishments of the company from 1979 to 1995. I have been affiliated with many technology companies over the last 30 years, and my experience with the original Network Solutions still stands out as one of the highlights of my career in the technology sector.

This was a company with a 16-year track record that garnered the respect and admiration of its customers: federal agencies and major commercial clients. The thing that truly differentiated NSI from other minority and non-minority firms was their in-depth knowledge and ability to solve highly technical problems in the communications area. During this period of growth, very few companies were experienced in linking mainframe computers to desktops.

At the height of Network Solutions' illustrious history, they had more than 430 employees and their annual revenue reached $48 million. This type of business performance over its history was a credit to the leadership of Emmit McHenry and his partners, senior management, and employees. The company had a network of offices throughout the country with its headquarters in Herndon, Virginia, and offices in Oklahoma City, Oklahoma; Chicago, Illinois; Los Angeles, California; Mt. Olive, New Jersey; Sacramento, California; Houston, Texas; and Detroit, Michigan

Hundreds of employment applications were received weekly from candidates across the country desiring to work for NSI and become a member of this highly recognized and skilled team. The company strongly supported diversity at all levels of its management structure which was not the norm with many other minority-owned technology companies during this period. Many of these companies believed that it was better to have a company that was predominantly white in order to attract more federal and commercial business. Instead, Network Solutions had a highly diversified team from its founders to the members of its staff. This diversity provided crucial experience and training for many managers and employees, who later created their own successful businesses following the sale of Network Solutions. Some former Network Solutions employees who started their own businesses are Karyn Trader Leight, Johns Marshall, James Walker, Scott Williamson, Andre Lynch, and many more.

Network Solutions typically provided services under multiyear contracts in three areas: cost-reimbursement (30 percent of total revenue), time and materials (60 percent), or fixed-price (10 percent). The last several years prior to the sale Network Solutions had more than a $60 million contract backlog (the unbooked portion of their contracts' future years).

Some of the outstanding accomplishments/clients in the federal and commercial space were:

- Air Force Headquarters Electronic System Division (AFSC)
- International Business Machines (IBM) in support of the Mission Operations Support Contract (MOSC) at NASA's Johnson Space Center
- Federal Energy Regulatory Commission (FERC)

- U.S. Department of Health and Human Services/Alcohol, Drug, and Mental Health Administration

- Air Force Systems Command Ballistic System Division (BSD)

- Defense Information Systems Agency (DISA)

In Contra Costa County, California, Network Solutions in partnership with the then known Pacific Bell, designed and installed an advanced digital network using technology at that time X.25 and TCP/IP packet-switching approach to solve multi-vendor interoperability problems and to provide high-speed end to end communication.

IBM/Mission Operations Support Contract: Network Solutions provided the systems software maintenance and integration support to IBM in support of MOSC at Johnson Space Center, including systems programming, engineering, and telecommunications.

Federal Energy Regulatory Commission: Network Solutions integrated existing microcomputers and mainframes located in FERC headquarters and regional offices in 16 cities using token ring LAN and using OS/2 LAN servers and IBM state-of-the-art technology. Network Solutions was the leader of this application in the country.

National Science Foundation: Network Solutions had been working with the NSF since 1987, five years prior to signing the contract to establish the Internet addressing system. They supported NSF large-scale advanced data processing (ADP) requirements with a focus on systems software support. The support included hardware selection, installation, and maintenance development, and management of the operations related to all mainframe systems: tuning/balancing, monitoring, programming, communications analysis, and support.

Government System Inc.: Network Solutions was the subcontractor to GSI under contract with the Defense Information Systems

Agency where they provided Internet addresses related to the Department of Defense's version of the Internet. The scope of the contract was to transition services to a Data Network Information Center (DDN-NIC) and Internet user. These services included network/user registration (network number and top domain name assignment), online information services and help desk operations. The work under this contract provided NSI with the past performance to later win a contract to launch the global Internet. No other company in the world had this breadth of experience and capability.

Sentinel Electronic Aspen System Division (ESD) USAF: Network Solutions provided systems integration and production support for the Air Force General's Applications Intelligence Training System. This support included hardware and software acquisition, integration, installing, and testing state-of-the-art, computer-based training platforms, and the associated network security operating system for various levels.

Nations Bank (Bank of America): Network Solutions provided the support to enable Bank of America to integrate the back-office operations of these merged institutions, ensuring a smooth transition.

These contracts represent just a sample of Network Solutions' capabilities during the '80s and '90s and were a testament to the company's technology expertise and experience at that time. These capabilities resided in the experiences and backgrounds of their management team and employees. Even though Network Solutions was considered a small to midsized company, they provided services to their customers as if they were a much larger enterprise. Many of their technologies were revolutionary for that time: integrating networks, security protection, interoperability solutions, telecommunications services and managing hardware mainframes or microcomputers, were the tools Network Solutions and its team used to solve customer challenges.

Al White Joins Network Solutions

During the George H. Bush Presidential Inaugural in 1989, I was the director of the African American Inaugural Committee as a registered Democrat. I was appointed to this position by the late Lee Atwater, who was then the head of the Republican Party. This appointment came as a big surprise to me and my family. However, Jimmy Whitehead, a close friend of Lee Atwater, convinced him that I was the right person for the inaugural position, even though I was a Democrat. Jimmy told Lee that it would help bridge the political gap by having a Democrat manage Republican inaugural events for African Americans.

This was not an easy task since most of the people on my team thought that my position should have gone to someone from their own party. The co-chairs for my segment were Willie Leftwich, noted Washington, DC attorney and Josue Smith, a successful businessman, who also was an established African American Republican. They protected me even though many loyal African American Republicans questioned why I was appointed the director of the Inaugural Committee and (they always let me know it).

We had the most successful inaugural event and President Bush chose our prayer breakfast event as his first inaugural event to attend. General Colin Powell was a special guest at our gala. We were all amazed that with all the other activities going on in Washington, that he stayed and danced throughout the night at our event. I learned a lot from the Republicans on how they raised political contributions and their link to major corporations in America.

Following the festivities, since I was not a Republican the possibility of a political appointment under the Bush Administration was remote. Emmit and I met following this experience and he wanted to know if I would consider working for Network Solutions. He knew that with my inaugural position I had developed relationships and contacts with some major political players in the Bush Administration that could be valuable in generating additional federal opportunities for NSI. Also, I needed a job. He announced to the company that he was hiring his friend to assist in the marketing and finance area.

In 1989, I joined NSI as a consultant to the chief financial officer (CFO), Roger Evans and assistant to Emmit McHenry. My becoming a member of the Network Solutions team was not well received by the partners nor by the other managers. They saw me coming to the company as Emmit giving a friend with no technology background a job.

In fact, Emmit had to hold an all-hands meeting with his partners and staff concerning my joining the company. Most people at Network Solutions except for Emmit did not know about my success in New York as a banking officer and owner of two companies. Additionally, unaware of my being an advisor to several noted African American businessmen, like the late Percy Sutton, the owner of the largest network of minority-owned radio stations in the country and my work with the federal government and state municipalities. Along with an extensive consulting background while at Columbia University and the University of Denver.

After the meeting, I was a member of the Network Solutions family as a consultant and under the watchful eye of the partners and staff. Partner Gary Desler asked me what business area I wanted to pursue. I told him that I thought focusing on developing a partnership with AT&T would be a real opportunity for Network Solutions. He broke into a huge smile, "Like really." He mentioned they had tried to market their services earlier to AT&T with no success. But the next question that he asked was how much help I would need in this effort. I told him it was not necessary to provide me with any resources (initially), but when I needed help, I would call upon him and the other partners.

Network Solutions Achieves
Recognition with AT&T

My prior entrepreneurial experience in New York and abroad had prepared me well for the task at hand. I had generally worked solo in my two companies and provided myself most of the operating services from marketing, finance, administrative, and contracting. When you work mainly on your own with limited resources, you learn to do whatever is necessary to stay alive and to thrive. So, executing a marketing and business development program by myself was not a problem. I just needed a well-conceived strategy and some luck. My experience as a former basketball player in high school and college provided me with the personal tools I needed—persistence and focus—to achieve whatever goal I desired.

When I joined Network Solutions, the General Service Administration (GSA), a federal agency, was preparing to award a historic contract to two companies for $50 billion. FTS 2000 was the largest telecommunications contract in the world. The contract was to manage the federal telecommunications services for the federal government: voice, data, and video transmission to more than 10,000 U.S. and foreign locations. They awarded the prime contracts to AT&T and Sprint. Each company would be enlisting the service of subcontractors to work on this 10-year project.

So, my strategy was clearly to win a subcontractor position for Network Solutions with AT&T. Since my wife, Gwen, worked for the federal group at AT&T heading up the Veteran Affairs account, I had a slight advantage of knowing who the players were at AT&T. Beyond that advantage, I had to do the rest on my own. One advantage of working on my own: I did not have to consult with anyone else on my strategy. But it helped to know that if you wanted to win a contract in any area, you needed to know the decision-makers and they needed to know you.

I arranged a meeting with Clyde Jackson, the head of federal procurement at AT&T. He suggested that I attend the pre-bidders conference they were hosting for potential partners on the recently won FTS 2000 contract. Clyde mentioned that I should stay in touch with his staff, Stephanie McNeil-Bates and Tanya Wilson, who would be reviewing the capabilities of each bidder. I obtained a copy of the RFP (request for proposal) and reviewed the requirements thoroughly before the pre-bid conference. The document stated that the contractor, AT&T, should consider minority suppliers at all levels of the work.

The bidder's conference was packed with major telecommunications companies who had lost the award to AT&T and Sprint, as the prime contractor. Additionally, there were numerous small businesses attending. I was thinking to myself, "What chance do we have against these Fortune 500 companies with more resources and capabilities than Network Solutions, for a share of the largest telecommunications contract in the world?" However, from my basketball experience, the size of your opponent is not important if you have a well-thought-out strategy and plan to enable you to score the critical "points" to win.

I discussed the opportunity with Emmit and his partners and received their approval to proceed and bid for the contract. I mentioned to them that we were facing competition from major telecommunications companies. For each opportunity, Network Solutions and other contractors submitting bids had to decide beforehand on the financial costs and resources involved in going after the opportunity. The resources to write a proposal and pricing tables, as well as any exhibits, had to be decided when there were limited resources.

I did not know the capabilities of Network Solutions' team to write a proposal of this type, but Emmit told me not to worry; his proposal team was very good and understood the opportunity. The final RFP had not yet been released, so I had some time to market Network Solutions' capabilities to some key people at AT&T and on Capitol Hill. Under federal regulations, I was not allowed to speak with the contracting staff of the offeror after the final RFP had been issued. I could only submit written questions to AT&T and receive answers to the questions along with all the other bidders.

I set up a schedule so that I could use every available day to meet with the decision-makers at AT&T. I lived in Silver Spring, Maryland, not far from their federal group headquarters office. Clyde Jackson introduced me to Dick Lombardi, the head of AT&T's federal division and his staff responsible for reviewing the bids. At every meeting, I talked about our capability of handling large and national contracts. I provided a written list of our past performance summaries and letters of commendation from customers. This information was to be included in the proposal, but it never hurt to emphasize NSI's success record early.

There were several individuals at AT&T who wanted the award to be made to a large telecommunications company to ensure the quality of the work. The contract required the winner to have the capability and equipment to do the work in all 50 states and abroad in providing data, voice, and video service to 43 federal agencies. I continued to market to AT&T at all levels knowing that Network Solutions was facing some tough competition and internal politics.

During my research of the decision-makers, I realized that the Congressional Government Operations Committee, headed by the late Congressman John Conyers was very important. He was responsible for monitoring the operation of the General Service Agency (GSA), the entity with the responsibility of managing the FTS 2000 contract for the federal government. Before the federal government instituted a security credentials requirement, I would show up in Congressman John Conyers' office and regularly meet with his staff: Ron Stroman, his chief of staff and Cheryl Phelps, a member of his administration.

At every opportunity, I talked about the outstanding capability and track record of Network Solutions. Emmit would attend several of these meetings in Conyers' office, I wanted them to meet the leader of the company and to hear of NSI's outstanding record. These informal meetings in the congressman's office continued even after the RFP was issued since regulations did not restrict our meeting with government employees not involved in the contract process.

To further cement my relationship with Congressman Conyers, I traveled to his district offices in Detroit and walked the streets with him. I was now known by his personnel in his Washington, DC, and district offices and they all were aware of Network Solutions and our capabilities. Many opportunities have been lost because bidders have a great proposal and competitive pricing but had very few internal supporters who knew the company or its capability before they submitted their bid.

Relationship building and showing your capability before the review process takes place can be the difference in winning or losing a contract. I was sure that I had built relationships with all the decision-makers and that they knew the capability of Network Solutions before the contract was awarded. The other thing that I had to take in consideration was the question, "Does a minority-owned company have the capability to handle a contract of this size and provide the breadth of services required?"

One factor that assisted me in my marketing effort was my physical size; being 6'5" and over 200 lbs., very few people denied me access to their offices. I was such a regular visitor at AT&T and Congressman Conyers' office, that those who did not know me but saw me regularly thought I was a marketing manager with AT&T or a member of Congressman Conyers' staff. Also, my upbeat, persistent attitude was something that I developed as a former athlete.

At last, the final RFP was released, and the Network Solutions proposal and pricing team started their work. I stopped calling on AT&T after the RFP was released but continued to communicate with Congressman Conyers' staff. On one of my trips to his office, I had the opportunity to ask the congressman what he would like to see as a result of the FTS 2000 opportunity. He immediately stated his desire to see minority-owned businesses get a share of the work. He felt that these businesses had traditionally been overlooked with opportunities of this size. He made his wishes known to GSA's small and minority business staff.

Network Solutions' bid and proposal team finished their work and I was tasked to deliver the box of proposals to Clyde Jackson's AT&T office in Silver Spring. When I arrived, my heart almost stopped when I saw 30 to 40 boxes of proposals sitting on his conference table, ready to be opened and reviewed over the next several days by his staff and the technical team. I should have anticipated this response due to the size of the opportunity.

Waiting to know if we had been successful in winning this major opportunity was like a mother carrying a baby for nine months. At least the mother knows there is some reward after the nine months. The difference in these two scenarios was that if I did not deliver the win, I would have been very depressed due to the time spent working on the project, but if we won, then the result of my work would be rewarded with a "baby," in other words a contract.

I felt that we had done everything possible to win this opportunity. But until we were officially told, nothing was guaranteed. When we did not receive a call on the designated selection date of the award, I became very nervous and received questions from everyone in the company. Why were they taking so long to announce a winner? I called Clyde Jackson's office to see if they had decided anything. He mentioned a large number of proposals that they had been reviewing and he would get back to me by the close of the day.

My mind was racing. Was the news good or was it bad? Finally, he called and asked me to come to his office the next morning. I could not sleep wondering what he was going to say to me. I drove to his office early the next morning and was not greeted with the smiles and eye contact I was used to receiving. I was really worried about what he was going to say. Sometimes these meetings were to let you know that they had awarded a contract to another company, but you came in second with a good proposal. I really did not want to hear that message.

He asked me into his office and stated we were a finalist with two other companies and the review committee wanted us to submit a best and final offer or BAFO. This type of request means that the finalist had questions around pricing and wanted the best and final price. He handed me the questions for us to answer within 24 hours and then they would make their final decision on the winner.

I called Emmit and relayed to him that we were a finalist and they wanted a best and final bid within 24 hours. All I could think about was another sleepless night without knowing. At least we'd made it to the finals, but this was little consolation for the work of the proposal team.

It all reminded me of a basketball game that my high school team lost on national television in the last few seconds. We were leading the game from the beginning and I was the high scorer for Erasmus Hall Boys High, a nationally ranked team was down by one point with three seconds to go and the star of their team, Eldridge Webb, threw up a shot at the buzzer to give them the win. It made no difference that we were winning the game up to that point; we did not win the game. However, the only consolation was that the television transmission ended with several minutes to go, due to another scheduled program and everyone except those who attended the game thought we had won. That was a feeling I never wanted to have again—being close, but not winning.

The next morning, the new pricing was delivered to Clyde Jackson's office and we again awaited their decision. I did not go into the office since I had very little sleep for the past two nights. But at noon, I received a call from Emmit that we had won the contract. I could hear the cheering and celebration in the background! I called Clyde to confirm the news and thanked him and his team. He told me that Stephanie McNeil-Bates would be working on a draft of the press release and would send it to me for review. I called Stephanie and she stated that the release would appear in all the major newspapers under the management of their PR department. I thought that was unusual for a subcontractor award. I went to the office and everyone congratulated me on the win, even some of the initial doubters among Emmit's partners. This validated Emmit's hiring decision of bringing me into the company.

When I received the draft of the press release from Stephanie, I was shocked to see, that it read: "AT&T has awarded its largest minority business contract ever to Network Solutions of Herndon, Virginia." It would be published on the cover of the *Wall Street Journal, New York Times, USA Today,* and other publications. The announcement was on all the major news shows for several days. This made Network Solutions a major name in the technology space in the '90s. Pictures were taken of Emmit and me with Clyde Jackson and Dennis Roth, the head of AT&T' FTS 2000 program. Network Solutions successfully performed at a 96 percent rating over the life of the contract and received several awards.

This started a long-term relationship with AT&T, working on the largest telecommunications contracts in the world. AT&T provided extensive resources and technical know-how to Network Solutions in the execution of the FTS2000 program. One of the federal programs established during this time was the federal government Mentor-Protégé Program. This program was developed to enhance the resources of small and minority businesses with the assistance of corporate partners. The Department of Defense and other agencies adopted this SBA program. They also assisted NSI with bridging some financial issues related to the contract. Their biggest support was their promotion of NSI and the references they provided, which enabled us to receive other contracting opportunities. Network Solutions' relationship with AT&T in the '90s became a model for other Fortune 500 corporations in their partnerships with small and minority businesses in the federal sector.

AT&T has traditionally supported the participation of minority and woman-owned companies even as they have grown and provided other communications services. Many small and women-owned companies have benefited from their relationship with AT&T.

Banking Dilemma Solved by the Win

One specific benefit of our winning the AT&T contract occurred a couple of weeks following the award. The Federal Reserve in 1990 closed many Washington, DC banks without notice because of questionable loans to board members. One of those banks was the National Bank of Washington, DC, the primary bank for Network Solutions. Emmit was told by our banking representative, "The Network Solutions account was closed until further notice like all the business accounts at the bank." Emmit told the banking officer that the company's payroll was due in two days and we needed some money to cover it. He responded that his hands were tied and that the person who was managing the bank for the Federal Reserve would not allow any withdrawals; several other companies had also requested some money from their accounts with no success.

Emmit called me to his office and told me about the situation and that Roger Evans, the CFO, was on vacation and would not return until the end of the following week. He mentioned to me that his bank account officer had suggested visiting the bank and speaking with the Federal Reserve manager controlling the operation of our branch. Emmit stated, "You're a former banker Al, you can work this out." No one in the company was told of the situation; not even the other partners.

We agreed that I should go to DC and see if I could speak to the Federal Reserve person about the dilemma; we needed to meet payroll on Friday. Emmit called back the manager of the bank and mentioned that I would be coming down and that he should introduce me to the Federal Reserve's manager so I could plead our case. The banker stated that no one had received approval to get their account opened until the following week when the federal audit would be completed. Luckily, he did get me into the bank and introduced me to the person designated by the Federal Reserve to oversee the branch's activities during this period.

I arrived at the bank at 10 a.m. and met our banker at the door with a security guard stationed to keep people out. I was allowed in and our banker marched me directly to the Fed's monitor for the branch. I was not sure what I was going to say to convince him to open our account. He stated that no account could be opened until they completed their audit of the bank's books. While rummaging through my briefcase to find a business card, I noticed that I still had a copy of the AT&T award article from the *Wall Street Journal* of our FTS 2000 contract. I pulled out my business card, and then the article and made this statement to the Fed monitor after showing him the article, "If we can't make payroll on Friday there is a possibility that we will have to shut down the government's communications system; all the phones and data links would be temporarily stopped."

That got his immediate attention. He placed a call to the New York Federal Reserve director, told him our situation, and mentioned the article from the *Wall Street Journal*. He asked me to wait in another room while he received instructions from his manager in New York. When I saw him again, he said that the Network Solutions account was now being managed by the Federal Reserve of New York and they would make arrangements to get us some money immediately.

I called Emmit and gave him the good news and could hear his sigh of relief over the phone. Truthfully, I did not actually think we had the power to shut down the federal government's communications system, but at least my story got the Federal Reserve's attention to make something happen!

When Roger Evans returned from vacation, he asked if everything was all right because the first call, he received upon his arrival was from the Federal Reserve of New York asking him how much money he needed that day. I explained to him what happened the previous week and he stated his relief of not having to deal with that issue and thanked me for my help. AT&T played an even bigger role in the history of Network Solutions in the next several years.

Chapter 6: Why Do We Have Minority Business Programs?

For years, many people have asked why we have a separate program for the minority business community. Those people asking the question generally believed that we have a totally equitable economy, but that is the furthest thing from reality. Even though the minority community today represents more than 45 percent of the population in the United States, we participate in less than 5 percent of the business opportunities in the federal sector, which accounts for more than $700 billion annually. This problem has existed for many years with little focus on the issue and without some regulatory support at the federal, state, and local entities. If this problem continued it would plague our economy and weaken the economic health and job creation in the minority community.

In the late '60s and early '70s, the minority community—both business and political organizations—wanted something to be done to provide greater access to the economic pie. The minority community defined in this period were mainly African Americans, Hispanics, Asians, and Native Americans. Several other groups were later added to this definition. It was the African American community that led the fight for economic equality on the federal, state and local levels.

During that period, the federal government's budget for the purchase of goods and services was around $500 billion annually. Many of these contracts were being won by prime contractors like GE, IBM,

General Motors, Lockheed Martin, and many more without participation by the minority community. The minority business community was facing an uphill battle to gain access to these opportunities. Since many states and counties also received federal funding, this problem at the local level was also being ignored. Minority businesses were being left out and no one seemed to care. However, members of the small, but powerful Congressional Black Caucus (CBC) started to make this issue known on Capitol Hill. This happened during the annual legislative CBC weekend were African Americans from across the country come to Washington, DC to make their voices heard relating to the problems facing their communities. Many of the issues in the '70s were focused on health, education, and the lack of jobs. Eventually, these annual meetings started to address the economic inequality in the federal government contracting system and the lack of opportunities for African American businesspeople across the country. There were "Brain Trust" meetings of African American businesspeople from around the country during CBC weekend and many were unhappy. How could the federal government take their tax dollars and not provide opportunities for them to participate in these huge federal contracts? This was at a time when our communities were losing jobs and the local educational systems in many African American communities were starting to crumble.

In 1970, there was an African American man from the city of Baltimore that was elected to the U.S. Congress, Parren Mitchell. My first exposure to Congressman Mitchell was during Congressional Black Caucus Weekend and he was the head of the Economic Brain Trust meeting. The room was packed, and everyone seemed to want to tell their story of being discriminated against in seeking federal contracting opportunities. They wanted to know what the members of the CBC were going to do. Mitchell came from a prominent civil rights family and was the brother of Clarence Mitchell who was responsible for several major legal wins around discrimination of public places and institutions. The fire in the belly of Parren when he came to the Capitol was raging and he took on the problem of improving African Americans' access to federal contracting as his primary goal. During these sessions on Capitol Hill, black people would ask basic questions about contracting opportunities with the federal agencies and what could be done to allow more African Americans and other minorities to participate. Parren surrounded himself with some of the brightest and most talented young African American attorneys, who were also committed to making something happen. Major Clark, Singleton McAllister, and the late Tracey Pinson were a few of the members of his team of "tigers" ready to take the federal procurement regulations apart and create new legislation that ensured that minority businesses would have a fair share of the federal dollars. Following the Brain Trust meeting on the Hill, Parren would invite individuals back to the office with more to say about their issues, for intense discussions and development of specific strategies. During the '70s, '80s and the '90s, many successful African American firms received major federal contracts due to the work of Parren and his team.

He attached to President Carter's $4 billion Public Works Bill an amendment that compelled state, county, and municipal governments seeking federal grants to set aside 10 percent of the money to retain minority firms as contractors and subcontractors; $625 million (15 percent) going to legitimate minority firms. He later introduced legislation which in 1976, which became Public Law 95-507, that requires proposals from contractors to spell out their goals for awarding contracts to minority subcontractors. This law still exists today. Parren and his tiger team added thousands of jobs and economic wealth to our communities then and even today. Major minority firms, like Network

Solutions, were able to partner with large prime contractors because of Public Law 95-507. This law provided access to billions of dollars of contracting opportunities for minority businesses. His amendment to the $71 billion Surface Transportation Assistance Act of 1982, required a 10 percent set-aside for disadvantaged businesses.

During his time on the Hill, the late Congressman Parren Mitchell held several positions on important committees: Senior Member on the Housing Banking Committee, Finance and Urban Affairs Committee, Chairman of its Subcommittee on Domestic Monetary Policy and Chairman of the House Small Business Committee, he also fought to protect the SBA 8(a) program and other programs.

Whatever committee the congressman sat on, fighting against the economic inequality of minority businesses and their communities were his focal point. Congressman Parren Mitchell was considered the father for "Minority Business Equality." Even though he is not with us today, he has left a legacy for his tiger team and the others that work for Parren to emulate his commitment to our community. Congressman Mitchell initially focused on opportunities for the African American business community then, other minority business groups today have benefited from his work (women, Native Americans, veterans, and others). They should be grateful to the tireless work of the late Congressman Parren Mitchell and the Minority Business Enterprise Legal Defense and Education Fund (MBELDEF), that he founded. One of his tiger team members, Anthony Robinson continues to promote the accomplishments of Congressman Mitchell as director of the foundations and annual award event to salute individuals that demonstrated their commitment to the minority business community or related to social issues in our community.

Following Congressman Mitchell's retirement, other members of the Congressional Black Caucus have continued to fight for economic equality for minority businesses and women-owned businesses.

Chapter 7: The Historical Event and the National Science Foundation

On December 31, 1992, Network Solutions under African American ownership, signed a contract with the National Science Foundation (NSF) to provide Internet addresses for the first time to the global community. This was an exclusive, five-year contract to provide Internet addresses and control of the Internet. No other entity in the world had ever been tasked with this responsibility.

The NSF is a United States government agency established under Public Law 570 in 1950 under President Truman. Its stated mission at the time was to promote the progress of science; to advance the national health, prosperity, and welfare; and to secure national defense. Its mandate was to provide grants to educational institutions and research labs across the country. During its history, the NSF mission has changed to include developing networks for colleges and universities and supporting the development of supercomputers. The NSF established the National Science Foundation Network (NSFNET) to provide a communications network for non-commercial entities before 1991. Then there were more than 5,000 networks comprising the Internet. The domestic, non-military portion of the Internet included NSFNET.

This was the framework for permitting the NSF to expand its service to the public. There was public pressure by the Clinton Administration to expand the use of the Internet to the public.

NSFNET was starting to be viewed by the Department of Defense community and other Internet entities as the only network capable of providing communication to the public. It was proposed that NSF take on the deployment of a non-military network program in 1991 to permit the commercial use of the Internet. This official change of support was with the assistance of Senator Al Gore's, getting the legislation passed: High-Performance Computing and Communications Act Public Law 102-194. Under this public law, the National Research and Education Network (NREN) allowed NSF to establish a structure to provide public access to the Internet. In 1991, the NSFNET linked together 16 different sites and more than 5,000 different networks. Many considered the birth of the commercial Internet as we know it today as happening in March 1991. "Senator Al Gore, the first political leader to recognize the importance of the Internet, promoted and supported its development," quoted by Vint Cert.

Since the National Science Foundation did not have the capability to manage the creation of the Global Network. They agreed to contract the actual operation and control of the Internet to outside contractors.

Global Internet Services Contract Award

The National Science Foundation in cooperation with the Internet community released Project Solicitation No.NDF24-92. There were three awards made in December 1992 under this opportunity: one to AT&T to provide database and directory services, as a directory of resources on the Internet; General Atomics, to provide seamless interface for Internet users; and the third winner with the most responsibility was Network Solutions, providing Internet addresses, for the first time to global users. Network Solutions created the means for the public to access the Internet or information superhighway (a term coined by the Clinton Administration). Specifically establishing the mechanism to issue Internet addresses for the public: individuals, companies, and organizations throughout the world, etc. .com,.org, net.

Having an Internet address allowed the public to have access to the Internet and the World Wide Web. These addresses on the Internet are like having a network of highways with vehicles: cars, trucks, and buses with a license to travel on the highway. This is the role that the Internet plays today by allowing the public to travel on the information superhighway.

Network Solutions Wins the Opportunity

Before 1993, when Network Solutions received the authority to issue addresses, only the military, educational institutions, and laboratories had access through an addressing system of the Department of Defense and the NSF. Network Solutions was instrumental in developing the early addressing system for the Department of Defense.

In an interview of the contract reviewers (stated in the book *Names, Numbers, and Network Solutions* written by Robert Beyster and Michael Daniel) the NSF evaluators awarded the contract to Network Solutions, stating that Network Solutions had extensive experience and strong past performance with similar requirements that were critical in NSI being awarded the contract. In 1991, Network Solutions got a glimpse of the future with the Network Information Center (NIC) that was transferred from Stanford Research International in Menlo Park, California to a company called Government System Inc. of Chantilly, Virginia. Network Solutions was retained as a subcontractor to oversee the Department of Defense's addressing system. There were few if any, companies in the world that had the experience of Network Solutions in this area. This contract was won by the company in open competition (any company was able to compete), but Network Solutions was the only bidder. No one at NSF was aware that Network Solutions was an African American-controlled company but recognized their experience and capability in this area. Also, all Networks Solutions customers had given the company high-performance ratings on other contracts that also played a role in them receiving the award.

When Network Solutions was initially tasked by NSF to issue Internet addresses, they were told they would be issuing these addresses for free, this is a very important point. The only funding provided NSI was under a five-year agreement—less than $1 million per year until 1998.

The specific tasks under Network Solutions' signed agreement of December 31, 1992, with the NSF were:

The Network Information Services Manager for Registration will provide registration services to the Internet community, register Internet entities such as autonomous system numbers and points of contact, manage the assignment of Internet numbers for the entire world, register domain names, manage the root domain for the Internet and provide help desk services to advise Internet users on the policies and procedural issues involved in Internet registrations, and provide information on such topics as domain naming schemes and Internet address assignments, as well as on the status of pending registration requests.

Domain names are Internet addresses, and the registration process creates a database that maps the names to the numbers used for Internet routing. When Network Solutions began operation in the spring of 1993, new domains were being registered at a rate of 400 per month. In October 1994, this number reached 2,000 per month and it was estimated that by the end of the year, the figure could top 20,000 per month.

Scott Williamson was the manager of the unit at Network Solutions with the responsibility of building and managing the first public distribution system for Internet addresses. He stated that "Initially the management at Network Solutions did not want to bid the contract for the work. However, I was able to convince the management that we had the experience and background." This was not a new area of expertise for Network Solutions, we developed a similar system, Defense Data Network(DDN) for the Department of Defense (DOD) in 1991, but unlike the DOD network where Network Solutions was issuing Internet addresses for a smaller number of people, the public Internet would eventually have millions of addresses issued during the first several years. No one had ever established an identification and registration platform of this magnitude, so it was new for Network Solutions as well as the National Science Foundation. No one foresaw the problems that Network Solutions would face in creating the first global addressing platform for the Internet.

One major issue after receiving the contract was the issuing of addresses to other than lawful owners of a name or brand. In fact, our monopoly over the issuance of Internet addresses was a problem in the early years. Numerous court actions were leveled against Network Solutions to limit our distribution of certain names that were trademarks of other companies. Like the type of problem that we face today when someone steals your identity and uses your social security card or driver's license or unlawfully gains access to your bank or credit card account. Many fast-thinking individuals wanted to register Internet addresses under the names of major corporations without having any direct association with these entities. Scott Williamson stated that corporations like McDonald's filed suits against Network Solutions for unlawfully issuing of e-mail addresses to people not associated with their company. Remember that these addresses could be acquired for free. When things are free there are no hindrances to acquiring something.

Since Network Solutions was the only entity designated to issue Internet addresses, most of the Fortune 500 companies and others that existed from 1993 to 1995 obtained their Internet addresses from Network Solutions. Even Amazon and Jeff Bezos received his initial Internet address from Network Solutions. Let me just list some of the well-known companies that possibly received their initial Internet addresses from Network Solutions: McDonald's, GM, Macy's, Sears, HP, Microsoft, Boeing, Ford, the *Wall Street Journal, New York Times,* NBC, CBS, and hundreds more. Foreign governments and foreign-owned companies also obtained addresses through our service. In the period between 1993 and 1995, more than 20,000 Internet addresses per month had been issued.

Another activity that took place during this period was when technologically savvy individuals who understood that these addresses could be very valuable in the future for businesses or services requested addresses like plumbing.com, electriccompany.com, help.com, run.com, safe.com, people.com, etc. They obtained these highly marketable names and sold them later for large sums of money.

One story in the book written by David Kushner, *The Player's Ball: A Genius, a Con Man, and the Secret History of the Internet's Rise,* is focused on the adventures of Gary Kremen, a Stanford University grad. In the early '90s, some people realized that certain names would be worth a fortune one day. Kushner in his book stated that Kremen scooped up some of the choicest names: job.com, housing.com, autos.com, match.com, and sex.com. In 1994, Kremen invested money into his newly purchased site, match.com which later became a multibillion-dollar business. I am sure you are asking what happened to sex.com. In Kushner's book, Kremen sold sex.com to a convicted con artist, Michael Cohen who turned the business deal into a multimillion-dollar enterprise with nine million members paying nearly $25 per month.

This practice continues today. Some of these Internet domain names have received offers of five to seven figures. During the first few years of operation, there was no limit on the number of addresses that could be acquired for free. This problem was later solved as well as the unlawful application for trademarked names. If you wanted to acquire the e-mail address of a major corporation's name or any other publicly known entity after the rule change you had to show that you owned the trademark and it was registered in your name. This requirement substantially reduced the unlawful acquisition of trademarked names.

Another costly aspect of our contract was the requirement to establish a 24/7 help desk. This would allow people anywhere in the world to ask questions about our service. Remember, we were issuing addresses for anyone on earth who wanted a free Internet address. We had to provide translation services for questions from non-English speaking countries. This was a very costly service and made the operation of our contract very expensive with no financial assistance from our client.

The cost of running this contract did not stop there. Each morning when the management team would arrive, they would see signs that hackers tried to break into the Internet address computer file. It's like what we see today with the problem of stealing credit card information and credit records. It was necessary for Scott and his team to acquire some sophisticated and costly firewall solutions (methods of stopping intruders). In the early '90s, this technology was still very new and there were few companies and products available to prevent a break-in on the master computers housing the Internet addresses. Hackers would try a new intrusion method every evening after we closed. We had to assume this cost of security protection, again, without any financial assistance.

The management of this operation became very expensive and the funding under the contract of less than $1 million annually was not nearly enough. This resulted in the founders taking drastic measures to keep the Internet running. The National Science Foundation was unable to increase the funding for the operation of the contract and suggested that Scott and his team file an application to request permission to charge a fee for the issuing of addresses to the global community. This application was filed around the second year of the contract (1994).

Remember, no one had ever set up an operation that required the issuing of hundreds and later millions of Internet addresses per day. Neither the NSF nor Network Solutions were aware of the problems they would face in setting up this global operation.

National Science Foundation
WHERE DISCOVERIES BEGIN

SEARCH 🔍

RESEARCH AREAS   FUNDING   AWARDS   DOCUMENT LIBRARY   NEWS   ABOUT NSF

Award Abstract #9218742

Network Information Services Manager(s) for NSFNET and the NREN

NSF Org:	CNS Division Of Computer and Network Systems
Initial Amendment Date:	December 31, 1992
Latest Amendment Date:	March 16, 1998
Award Number:	9218742
Award Instrument:	Cooperative Agreement
Program Manager:	Donald Mitchell CNS Division Of Computer and Network Systems CSE Direct For Computer & Info Scie & Enginr
Start Date:	January 1, 1993
End Date:	September 30, 1998 (Estimated)
Awarded Amount to Date:	$4,369,334.00
Investigator(s):	Deborah Fuller (Former Principal Investigator) R Williamson (Former Principal Investigator) Mark Kosters (Former Co-Principal Investigator) Robert McCollum (Former Co-Principal Investigator)
Sponsor:	Network Solutions Inc Vienna, VA 22180-
NSF Program(s):	ADVANCED NET INFRA & RSCH, APPLICATS OF ADVANCED TECHNOLS
Program Reference Code(s):	0000, 9180, 9217, 9251, HPCC, OTHR, SMET
Program Element Code(s):	4090, 7257

ABSTRACT

The Network Information Services Manager for Registration will provide registration services to the Internet community, register Internet entities such as autonomous system numbers and points of contact manage the assignment of Internet numbers for the entire world, register domain names, manage the root domain for the Internet and provide help desk services to advise Internet users on the policies and procedural issues involved in Internet registrations, and provide information on such topics as domain naming schemes and Internet address assignments, as well as on the status of pending registration requests. The project is one of three collaborative activities comprising the Network Information Services Management Project.

Please report errors in award information by writing to: awardsearch@nsf.gov.

Summary of Network Solutions Agreement with the National Science Foundation (NSF)

Chapter 8: Selling an Invisible Concept

Following the signing of Network Solutions' five-year contract with the National Science Foundation, my role as vice president of corporate communications was to promote this service to the global community. No other member of the Network Solutions management team was tasked to promote this new service.

During the Clinton Administration, the generic name for the Internet was the "information superhighway". This was proposed as a high-speed communications system that was touted by the Clinton/ Gore Administration to enhance education and business in America in the 20th and 21st centuries. Its purpose was to help all citizens, regardless of their income level, gender, or ethnic identity. The Internet became the information superhighway whether we were ready for it or not. There were also other major initiatives to promote the use of the information superhighway for business use.

Since the information superhighway was not widely known or understood by the public, my assignment seemed almost impossible. There were very few people who knew why they needed another address. They already had a mailing address that they used on their letters and packages. What was the purpose of this new address? Since Network Solutions was offering Internet addresses for free, many people believed they could take a chance on "riding this electronic highway" by acquiring one or more addresses. They still were not sure why they needed this additional address.

Even in 1994, Bryant Gumbel joked with Katie Couric on NBC's *Today Show* about not knowing what the Internet was. They even thought that the @ sign was an abbreviation for where to go, "But where?" he asked. They were not totally wrong; it was clear they did not understand that the @ sign linked the user to a global network of computers that was an electronic post office delivering or retrieving information. Click on the link below to listen to their discussion.

https://www.youtube.com/watch?v=95-yZ-31j9A

Many technologists, including Bill Gates, did not initially accept the information superhighway as a technology platform with any real potential. If the founder of one of the most successful technology companies in the world did not believe the Internet was anything more than a good idea developed by the U.S. Department of Defense, imagine what a daunting task it was for me to sell the public on the concept. Nonetheless, I began putting together my marketing program.

I personally worked with my operations team issuing the Internet addresses and assisted in creating the first marketing materials with my assistant, Mary Block. I spent considerable time researching and meeting with individuals in the federal sector and NSF to understand their approaches in marketing this type of service to their customers. No one had any answers on how I should market this technology to the public because it had never been done before, but potential future use was projected to be enormous—the world! Unlike the introduction of the cell phone and computer in the '60s and '70s, most people knew how they would benefit from the use of these technologies.

In my presentation, I had to first describe what the Internet was and its history. Then, I would proceed with information on future uses by the public. I used PowerPoint to develop this presentation, not realizing at the time that this type of packaged software in a box would be extinct once Internet technology was fully deployed. We can now download Microsoft products from the Internet.

In the presentation, I discussed how the general public could utilize this platform once they had their address, a computer, and a link to this global network. My presentation gave a simple description of the future services and opportunities in this new technology platform. One example that I provided of how people in the future could use the Internet from home to select the color and design of their loved one's casket. This example always got some chuckles from my audiences. It made an important point about the capability of the internet, being able to make purchases without visiting a showroom.

I made multiple presentations at numerous conferences, events, and companies that were interested in educating their staff on this new technology. I even represented Network Solutions at Super Bowl events in Miami. Moses Brewer of Coors Brewing Company asked me to do a presentation to his major beer distributors attending the Super Bowl. At many of these events, the audiences were still unsure about the value of the information superhighway and its potential. Following many of my presentations, I received a few questions, but most people would simply ask about access, when it would be available, and how to acquire an address to use these services. I could see in the eyes of my audience that they were all trying to comprehend the information that I was presenting to them and asking themselves, "How can this be possible?" On many occasions, I felt like a religious leader telling the audience that a new day was coming, and their lives would change, but they needed to accept this technology path to heaven. Unlike a real religious leader, I could not promise them eternal life if they traveled down this path.

The following is an excerpt from my presentation in 1994:

"The information superhighway or the National Information Infrastructure (NII), connotes the means of providing information on fiberoptic or wireless networks, throughout the country—in your home, school, car—anywhere a person desires to receive information. No longer would you need to go out to get a newspaper or magazine, travel to crowded malls to shop for clothes, appliances, gifts, or to the video store to pick up a movie or video game. All these services will soon be provided on your computer, wireless notepad, and even your television. Information of all types will be available on DEMAND."

During this period, I received a call from IBM's marketing department asking me to participate in a program called PATHWAY TO GROWTH. Since Network Solutions was the company responsible for providing the addresses to the information superhighway, and I was responsible for leading the marketing effort, I was asked to participate in this national program to promote the information superhighway and its potential for businesses. IBM planned to invite local business-people (mainly their customers) from 10 major cities to hear about the coming opportunities. I was a speaker at several of these conferences and provided information on how to obtain an address.

Other presenters on the PATHWAY TO GROWTH program were potential vendors of products or services on the Internet. The 1-800-Flowers president, real estate owners, healthcare professionals, and others discussed the types of opportunities that would be available on this new communications platform. Their focus was on promoting the capability of this technology and the ease it presented to their customers of buying their product or service without having to leave their home or office. Audiences were amazed to hear that eventually; they could buy their products and services without going to the store or mall. This was the first time I had heard this type of presentation about the Internet, but I realized what was driving this initial interest from the other presenters were *business opportunities*. In addition to the IBM program, I received invitations to speak at federal and state government events for political officials and their constituents.

I was also starting to see books and research reports on the growth and future opportunities on the information superhighway. One of the research reports that I found in 1994 was produced by the Morgan Stanley Company, which stated that opportunities on the information superhighway would grow to $3.5 trillion by 2000. At the time, this seemed completely unbelievable to me since the public knew very little about this technology and how they could use it. However, quicker than we imagined the demand for Internet addresses grew from a few hundred per month to more than thousands of addresses issued monthly in few years eventually millions.

The overall volume of requests for information increased each day. Some other questions included were, "Where is this information superhighway?" "Can I bring my children, and do they need their own address?" "How much would it cost to ride on this highway, and does it go through my town?" One additional aspect I needed to explain was that in addition to an Internet address, they would need a computer and a telecommunications connection that linked to the global network of computers on the Internet. Today, there are Wi-Fi networks everywhere and devices you carry in your pocket, but access to the Internet in the early '90s and 2000s was still not available everywhere or to everyone. Many rural and urban communities lacked adequate computer equipment, and access to the network was nonexistent for many years.

Before public access to the Internet, most people used their computers to write papers, play games, and to educate themselves only when they acquired software packages that were loaded onto their computers. The general population did not envision that these types of activities could be accessible with an address on the Internet and connected to the World Wide Web. It is difficult for young people today to comprehend that we had to retype documents repeatedly before Microsoft and Apple created software that could make corrections on the fly. Access to the Internet made it possible to research words on an electronic dictionary and access spell check applications. There was no longer a need for volumes of bulky encyclopedias to research facts and history, the Internet replaced this tedious and time-consuming process with just a "click."

Considering the range of what people are doing on the Internet today, I could not have imagined the vast amount of services that have been developed over the last 25 years utilizing the capability of the Internet. One of the fastest-growing industries on the Internet today is pornography. This industry has benefited greatly from the Internet's accessibility. I certainly never expected this to happen and doubted anyone else did, but those owners of these businesses who have made millions and today billions of dollars.

The financial institutions' projection was only a few years off on their estimates, but this technology gathered momentum once the public realized the power that the Internet could provide to communicate globally. Even now, this momentum will continue for decades with new applications being developed like the Internet of Things (IoT) and self-driving vehicles. The Internet has substantially reduced the time it takes for new technology to be developed and distributed to the public. By the year 2000, just seven years from its inception, more than 500 million Internet addresses had been issued. And there are literally millions of feasible ideas being generated by people all over the world using the Internet: payment systems, healthcare solutions, self-taught education programs, etc.

Since Network Solutions was a minority-owned company, I felt an obligation to provide as much information as possible on opportunities created by the early adoption of Internet technology and the development of businesses around the potential services. Unlike the presentations I gave to general audiences, I tailored my presentations to minority audiences around the ability to market products and services globally without racial identification. The information superhighway was color-blind. This could open numerous opportunities for minority businesses. I also wrote articles for the ethnic press and print media on the increasing opportunities on the Internet. Whenever I met with a family member or friend, I would provide an overview of this game-changing technology and the need to get on board early. But it was difficult for the minority community to fully comprehend this opportunity because they were dealing with so many social, political, and economic issues plaguing their families and communities in the '90s. When I meet people today who remember my presentations, they often say, "Al, we should have taken your advice and ventured into business opportunities on the information superhighway back then."

One of the major problems that impacted minority communities at that time, even if they had embraced the concept, was access to the information superhighway in their community. The ability to ride this technology train required a couple of other things I have mentioned before: a computer and access to the Internet through telecommunications networks. Gradually through federally funded programs and donations by corporations, these problems were partially solved for some, but today some rural areas and southern states still lack total Internet access 25 years later.

I was very disappointed that I was unable to convince and motivate more people in the minority community of future opportunities on the Internet. In fact, I sent a letter to the editor of a large minority business publication suggesting that they publish one of my articles on the opportunities for minorities on the Internet. He responded that they had no interest in promoting that type of technology in their magazine. I met the retired editor recently and he admitted he had made a big mistake in not publishing my article. If this article had been published, it could have promoted a better understanding of the business opportunities on the Internet in the '90s, maybe motivating some in the minority community to create businesses rather than just becoming major Internet consumers. Maybe we could have developed technologies like Facebook, Google, Uber, or Airbnb.

African American Trailblazers in the Early Adoption of Internet Technology

There were several individuals from our community who saw an opportunity on the information superhighway. One person was Omar Wasow, a college professor. He would call me at my office and quiz me on how he could obtain an Internet address as well as questions about how he could do business on the Internet. Omar mentioned that he lived in Brooklyn and since we have similar roots, I provided him with as much information available to me on Internet technology. Our discussions went on for several weeks concerning the growth in addresses being issued. Later Omar cofounded his own Internet company in the social media space called BlackPlanet.com, one of the first African American websites. He helped grow this company to three million users. Also, during this time Omar was a regular speaker on panels discussing what he had achieved on the Internet and promoting its use in our community. Several noted celebrities consulted him in the early days of the technology; one of his consulting clients was Oprah Winfrey. He later sold BlackPlanet.com and I assumed made a substantial amount of money.

Another person who sought my advice during the early '90s was the social and political activist Faye Anderson. She also grew up in Brooklyn and attended Stanford Law School. Faye wrote numerous articles on the information superhighway as well as published a newsletter on this new technology in 1994 and 1995. I contributed several articles to her newsletter that was primarily directed at the African American community. Faye was a regular guest on Bill Maher's early TV talk show, *Politically Incorrect* during the early days of the Internet.

Larry Irving worked for the late Ron Brown, the first African American Secretary of the U.S. Department of Commerce during the Clinton Administration. The President had appointed Secretary Brown the task of coordinating U.S. industry efforts to better understand and take advantage of the information superhighway technology. Larry was heading the group responsible for devising the initial business strategy for the business community. With two African Americans leading this effort, I was hopeful that our community would see the importance of this technology and become more involved. I was critical of the agency not providing greater support to our community in taking advantage of the business opportunities and not just focusing on major corporations. Finally, Larry Irving started to participate in small business conferences discussing why our community should embrace the information superhighway and its potential.

Other individuals from the African American community that were involved in the early days of the Internet, as stated in *The Root* article by Genetta M. Adams, "Black Internet Pioneers," February 29, 2012, were as follows:

- Mike Holman, creator of Holman's World BBS

- David Ellington who worked with Malcolm CasSelle to launch NetNoir, the first online service devoted to Afrocentric content in 1995

- Kwame Anthony and Henry Louis Gates Jr. who were coeditors of the CD-ROM Encarta Africana, the comprehensive collection on the diaspora, which Microsoft released in 1999

- Darlene Dash of Digital Mafia Entertainment Interactive Holdings, one of the first African American-owned Internet companies on a publicly traded (Nasdaq) stock exchange in 1994. She is a native of the Bronx, New York, whose sister is actress Stacey Dash and is the cousin of Damon Dash.

- Chinedu Echeruo, creator of Hopstop, a website and phone app that helps navigate major cities in the United States, Canada, and Europe

- Bonita Coleman and Cheryl Mayberry Mckissack in 2000 developed websites that catered to African American women as an offshoot of their print publications.

The other Internet pioneers listed in Adams' article were not influenced by the launch of early Internet development.

It was a pleasure to see some bright and energetic African Americans investigating and starting early enterprises in the space. The numbers increased slowly during the following years because of the lack of capital for startup operations even during the dot-com boom. This technology was still foreign to investors in our community. That has changed today, but our community is still not creating enough meaningful and scalable projects in the space. This needs to change to make our communities Internet powerhouses.

When we ignore technology because we do not understand its purpose, we are limiting our opportunities. Technology today is moving at lightning speed; you are not going to understand every new technology, and if you wait until the technology is fully adopted, it is generally too late to benefit financially. You are then relegated to the status as a consumer, not an owner.

Chapter 9: Facts about the Selling of the Original Network Solutions

One of the big questions that I have been asked over the years since I was a key member of the senior management team at Network Solutions in 1995 when the company was sold is: Why did they sell the company? How could you sell a company that was the platform for the third economic revolution? Even within my own family, my older brother, Martin, asked me with astonishment in 1995, why would the owners sell the keys to control the Internet?

In 1995, the Internet was becoming a popular technology in various circles. The Clinton Administration had designated the information superhighway as one of the most important technologies of our time. There were substantial funds being allocated to support the building of the infrastructure to support this information superhighway that would change the way we communicate forever. During these times, why did the owners of Network Solutions sell the company?

Less than three years after NSF awarded Network Solutions the right to provide Internet addresses to the global community (December 1992), the owners sold Network Solutions. Through the services of Network Solutions, the global community had the opportunity to travel the information superhighway and be linked to the world.

The sale of Network Solutions at that point in its growth (and on the verge of the explosion of the Internet) was questioned by many inside and outside the company. The late Ron Brown, head of the U.S.

Department of Commerce considered access to information as the primary tool for economic growth.

The decision to sell was under the control of the four founding partners, Emmit McHenry, Tyrone Grigsby, Gary Desler, and Ed Peters. Michael Daniels, a VP of SAIC (an employee-owned federal contractor) had been courting the partners and tracking the growth of the company for several years. The actual sale of the company to SAIC did not take place until three years following the signing of their contract with the National Science Foundation; Network Solutions had revenue of more than $35 million and 375 employees around the country. Michael Daniels held meetings with Emmit McHenry and other partners for months, and they had turned down several initial proposals to sell.

Historically, NSI had faced a couple of rocky years of trying to generate profits to support its growth. Several years of negative earnings had placed a strain on the company's owners and employees. When they received the NSF contract award, they were unsure of the financial risk of being the first company in the world to issue Internet addresses. As already mentioned, the five-year contract provided a little less than $1 million per year and NSI was offering the public Internet addresses for free. This was a major problem because as the number of addresses issued was increasing; so, did operating costs. I must make the point again: these addresses were being issued for free.

It has been already stated: there were extensive costs related to the management and operation of the Internet addressing system: having to assume the legal cost of suits related to issuing of unauthorized addresses to companies without official trademarks; establishing a 24/7 help desk to answer questions from all over the world; providing translation services; numerous suits from corporations and individuals; protecting the addressing system computers from illegal intrusion by hackers. The National Science Foundation was not assuming any of these additional costs. These unanticipated expenses pushed the company into a financially difficult position. Even their other NSI contracts were suffering due to the lack of capital that was being redirected to keep the Internet up and running.

In 1993, the first year of the operation under the NSF contract, Network Solutions sustained financial losses of $450,000. In the second year, more addresses were issued, and the financial losses reached $950,000.

The owners of Network Solutions made a request to the NSF to permit them to charge a fee of $50 for each address. This request languished in the offices of NSF for almost two years from the date submitted and was finally approved several months following the sale of the company to SAIC. If this approval had been awarded sooner, it would have substantially reduced their losses and put the company in an improved cash position. Other financial issues faced NSI, the acquisition of a technology company in Florida that did go well and cost the company a substantial amount of cash. They were unable to integrate the technology of the acquired company into their operation successfully.

With losses mounting from the NSF contract and personal financial issues facing the owner Emmit and his partners, he sought investors from the African American community and the majority community. Emmit had meetings with members of the Presidents Roundtable, a group of minority-owned businesses operating in the federal market. Additionally, he met with a successful African American businessman in the entertainment industry and other groups of potential investors. In several of the meetings that I attended with Emmit, the responses from these groups were mostly negative. Many of them did not believe that an African American company was given such an important role of being the first company issuing addresses for the global Internet community. Some did not even understand the importance of the Internet and doubted the value of our role. We faced a similar response from the Wall Street community and since we could not charge a fee for issuing addresses, they could not see the return for investing in Network Solutions. In addition, we also faced a low-level of belief from the white firms that an African American company would have such an important role. The white partners also participated in a number of presentations. In a couple of cases where there was some interest from businesses that understood the importance of our role, they were not moving quickly enough to decide, maybe intentionally hoping to get a "fire sale" price for their investment. Most knew the company was in a desperate financial situation. The biggest disappointment for Emmit, Ty, and myself was the lack of support from the African American business community and its lack of respect in what we were providing globally. I hate to say that after more than 20 years, there still is little support among African American businesses to partner, merge, or invest in companies within our communities. This problem restricts our economic growth today. Even if Network Solutions had founded Facebook or Google, I am not sure we would have been able to raise the capital from our community to promote the operation of these companies.

I was concerned that the founders would make a premature decision concerning NSI's future and I solicited the assistance of our long-term partner, AT&T, in providing financial assistance. At the time, AT&T had a venture group that invested in technology companies. Since we had a close relationship with AT&T, I called Clyde Jackson, our major supporter at the company. He contacted the treasurer of AT&T who recommended that I present our case and our specific financial need to their venture capital group. I supplied all the necessary financials for their review. I stated that we required at least $5 million to support Network Solutions' operation work under the NSF contract and that we had been operating for the last two years. AT&T was aware of our work since they also had a contract with NSF providing supporting services under the same master agreement in a different area. The head of the venture group called me and said that they could provide the funds I requested, but all AT&T's funding needed to go into the company. Their review of NSI's balance sheet showed several extensive loans to the owners and AT&T wanted all their money to be invested in the company and not go to the owners. I mentioned AT&T's conditions to Emmit, and he stated that they could not accept AT&T's funding offer because the owners needed to take money out of the company to cover some personal needs of one or more partners. I thanked AT&T for their assistance and told them that the company would have to look for other alternatives.

The most viable option that could solve the personal needs of the partners appeared to be the sale of the company.

The Acquirer

Science Applications International Corporation (SAIC) was not widely known outside the federal and state contract arena in the '90s. It was founded in 1969 in La Jolla, California by the late Dr. J. Robert Beyster, renowned scientist and engineer in the nuclear science arena. Dr. Beyster, a former naval officer, worked at several federal installations as a nuclear scientist prior to establishing SAIC. The company was one of the largest employee-owned companies in the U.S. with more than 5,000 employees and sales exceeding $4 billion. Many of the SAIC management teams were retired military officers who brought with them an understanding of the various military commands and their influential contacts within the Washington, DC political community.

SAIC was a major provider of technology solutions and integration services in the healthcare arena for public and private entities. They also provided extensive expertise to the cybersecurity and intelligence community.

Michael Daniels was the person at SAIC who discovered Network Solutions. He grew up in Missouri and was a former Naval Reserve military officer. He spent considerable time with CACI, a major federal contractor before moving on to SAIC.

He sold his company to SAIC several years before the Network Solutions transaction. He became a senior vice president at SAIC, following the sale of his company. He had been courting the owners of Network Solutions since 1987 and continued to track the company until the sale.

Daniels did not initially recognize the NSI opportunity as a player in the Internet space, but for their successful track record as a provider of technology solutions to public and private entities.

Sensing that the sale was imminent, I made a last-ditch effort to persuade Emmit that there may be some other alternatives since I was seeing the acceptance of the Internet grow. I arranged a meeting with Emmit at a private club in Washington, DC, the Franklin Square Club. I had asked one of the NSI board members, General Arthur Gregg, to sit in the meeting as a third party, and act as an arbitrator. I proposed to Emmit that he and I buy out the three partners and we would continue to operate the company as partners. That proposal was not acceptable to him since the SAIC offer was on the table and the partners could close in a few days. Since I was not a partner but had contributed substantially to the growth of the company, I requested some equity shares. This was also rejected by Emmit and he was unwilling to ask the partners about my request. Our meeting ended, with me unable to change Emmit's mind relating to the sale of the company nor to obtain any equity for my contribution.

I was extremely depressed about the sale because I believed Network Solutions controlled the next major technology and if they waited, even if just another year, it would reward them an enormous return.

This sale officially took place in March 1995. Emmit was convinced to do a deal because SAIC agreed to take care of NSI employees and provided the partners with a financial structure that would meet their individual financial requirements. There was also discussion at a private dinner with Beyster and Daniel of Emmit becoming a board member of SAIC following the sale, but this never occurred.

The Terms of the Sale

On March 10, 1995, SAIC officially purchased all the shares of Network Solutions from the four owners. They received $5 million of employee-owned shares of SAIC that were not freely tradable outside of the company. Since these shares were not freely tradable stock, an outside security company was retained to provide a value for their shares related to their individual equity ownership in Network Solutions. There were several members of the management team that received some cash compensation or forgiveness of any company loans.

None of the owners held any Network Solutions shares following the sale. A hundred percent of the shares were now owned by SAIC. This fact would later cost the owners substantially.

One of the ironic things that happened following the sale of Network Solutions to SAIC, was the approval of Network Solutions applications to charge a fee of $50/$100 within six months following the sale of the company. Many people who have heard the story, questioned why this decision to charge a fee was made so soon after the sale of the company to SAIC. Why was the modification of Network Solutions' original application not approved until the ownership changed?

Once Network Solutions was owned by SAIC, they started charging a fee and generated revenue to offset any losses and finance the buildout of the infrastructure with the substantial increase in the number of Internet addresses issued.

Network Solutions as a Public Company

One major problem the four owners of Network Solutions had during their operation under the NSF agreement, was that they could not show any revenue projections from their contract since they were issuing addresses for free. This made it difficult for them to put a value on the company beyond their other business operations. Not until NSF approved the fee structure for issuing addresses of $50 for one year and $100 for two years, on September 15, 1995, was it possible to calculate a valuation of the contract with the National Science Foundation.

Within two years following the sale to SAIC, NSI was registered as a publicly owned company on September 26, 1997. The initial public offering (IPO) was structured for sale of 3.3 million shares of the Network Solutions stock. SAIC sold some of its shares under this $50 million offering and raised $9 million. This IPO and market price of $25 per share for Network Solutions stock translated into a public market value for the company at more than $382 million as stated by Michael Daniels of SAIC.

Now with an established value of the NSI stock, SAIC authorized a secondary public offering on February 8, 1999, of Network Solutions stock. NSI sold 4,580,000 shares of its Class A stock and SAIC sold nine million shares of their NSI stock. This offering raised $779 million, considered the highest value of an Internet company at that time.

Network Solutions' fortune continued to increase for years following the sale of the company to SAIC.

Chapter 10: After the Sale of Network Solutions

Each partner had to utilize their SAIC employee shares as collateral to borrow in receiving cash after the sale closed. SAIC also assumed at least $5 million in the short-term debt of NSI. Additionally, each partner (except Ty Grigsby) was offered a position in the newly constituted Network Solutions. Gary and Ed accepted positions and Emmit was under a consulting agreement until he established a new company, Netcom. Since SAIC could not assume the contracts that were based on minority ownership, Emmit acquired some of the Minority Business designated contracts for Netcom.

Mike Daniels, the person responsible for orchestrating the deal for SAIC, asked me to meet with him. He wanted to know if I would continue to work for Network Solutions, under the leadership of Emmit. I told him that I could not accept the position since I was still upset over the sale of the company. Additionally, I could no longer work for Emmit due to issues pertaining to the sale and our differences. Mike Daniels knew that I wanted to travel to Africa and to evaluate the possibility of SAIC developing a relationship with the management of a major new hospital and health center being built in South Africa. SAIC had been a major player in the healthcare arena working with federal and state health organizations. He agreed to my trip and suggested that George Otchere, a senior manager at SAIC accompany me on the trip.

George and I had known each other for a couple of years since he was responsible for the Small and Minority Partnership program at SAIC. He was responsible for identifying potential partners in the small and minority business community that would be potential partners on future federal and state opportunities. As mentioned, the federal government was supporting the partnership of major corporations that partnered with small and minority companies.

This was not the first time that I had traveled to Africa. While an international banking officer and consultant, I made a number of trips to the continent. Also, as a consultant to the late Percy Sutton, we traveled to Africa to evaluate the possibility of Inner-City Broadcasting establishing a record pressing factory there in Nigeria. This was a passion for Sutton and every time we traveled to Nigeria, we would be greeted at the airport by large numbers of people who knew him and his broadcast empire. During our trip, we would tour various sites for the facility.

My role in the record pressing project was to raise funding. Since I had an extensive banking background in raising funds for multimillion-dollar international projects, he sought my assistance. The total project size was $14 million. It was $12 million of debt and $2 million of equity. I received a commitment from the Export-Import Bank of the U.S. for the debt, but Sutton was responsible for raising equity. At the time, he had provided extensive amounts of his personal capital to fund the radio stations. So, we sought to raise equity from outside investors. At this time, it was very difficult to get investors for projects in Nigeria. There had been some major cement delivery scandals and other questionable transactions. It was reported that billions of dollars were lost by investors around the world. After several months of diligently meeting with African American investors and bankers, we were unable to raise the equity required and the project died.

My other international work was as an advisor to the city of Miami, attempting to develop relationships with Africa and provide trade venues for the marketing of African products. During this period, the Hispanic community in Miami was flourishing as a result of trade links with Latin America which still exist today. The African American community in Miami, as well as Dade county commissioners who were predominantly African American, wanted to investigate the possibility of developing a similar link with African countries. My partner, Shirley Barnes, and I were awarded a contract to travel around Africa and investigate the potential of bringing African-made products to Miami for sale in the department stores and trade fairs. This was in the late '80s and a number of the African countries had just emerged as independent nations and had not been able to produce products to even satisfy their local market. Our evaluation found after visiting several countries and meeting with several African businessmen that they did not have the quantity or quality for consistent business opportunities in Miami or anywhere else. This has changed substantially over the years, with African-made products being sold all over the world, some businesspeople also utilizing the Internet to promote their products.

So, following my trip to South Africa as a consultant to SAIC and filing my report, I officially resigned from Network Solutions.

Moves to Another Minority Company

It did not take me long to find another position. I was unaware that there was another minority-owned company that had been tracking my career; when I was available, they hired me as senior vice president of corporate communications. This company was RMS Technologies Inc., headquartered in Marlton, New Jersey, with its federal contracting office in Lanham, Maryland. The company was founded by the late David Huggins, a federal contractor providing services to the FAA, NASA, and Prince Georges County. When I was hired by the company, they were doing more than $100 million in revenue with a substantial net income. They had more than 1,000 employees assigned to major FAA facilities and other federal installations around the country. They provided computer integration services and engineering support. I received a call from the late Rotan Lee, the executive senior vice president at RMS recently hired by David Huggins to oversee the commercial operations which did not exist. Rotan Lee was a member of the Philadelphia School Board and some private sector boards. Huggins hired Rotan because of his political contacts and his ability to negotiate difficult situations. He was a highly respected attorney in the Philadelphia community. Rotan wanted to meet with me and discuss my joining RMS Technologies to help build their commercial business. He made me a substantial offer to become their SVP of commercial business. I still did not know why they were interested in hiring me until after I had been with the company for several months. I asked Rotan. He stated that David Huggins had told him that, "there is one person that stopped us from becoming a larger company than we are today and that is Albert White. He stole the FTS 2000 contract from us and our relationship with AT&T." We had the contract in hand, we had inside people at AT&T supporting us and we had done a great proposal; we knew the contract was ours. Then we discovered from our inside contact at AT&T that Network Solutions won the contract even with all the work we had done. Offline, Huggins had asked his internal contact at AT&T, what happened. "You were out marketed by a guy named Al White," he was told. Network Solutions had the support of the internal managers at AT&T and the political support of the Government Operations Committee. RMS was the second-best but did not win the game. With that knowledge, Huggins was determined to hire me if I became available. He instructed Rotan to contact me when he learned that Network Solutions was sold. What is ironic about my relationship with RMS and Huggins is they were the only African American company prepared to make an offer to buy the original Network Solutions. Huggins had a deep understanding of technology and was aware of the potential of the Network Solutions contract with the National Science Foundation. He had been tracking the growth of the Internet and believed it was worth pursuing ownership of the company. Emmit and the other founders felt the terms being offered by Huggins were not acceptable.

So, I assumed my role as senior vice president of RMS Technologies and worked to develop a brand and image for the company beyond their federal and state government business. I made several suggestions on how they could improve their marketability with major corporations by focusing on their technical strength and their national capability. I developed a new line of marketing materials for RMS that focused on their size and technical capability for the commercial market.

During my first management meeting at the Marlton, New Jersey headquarters they discussed their concerns about the direction of the commercial marketing program. Since the company had little experience in the commercial arena, they were unwilling to listen to new ideas and someone with experience. This unwillingness to make the necessary changes in their strategy from a federal-focused strategy to commercial limited their growth in this area. The major difference between the two business areas was the expectation of each customer. Most federal business development takes a lot longer than the commercial market business development. Most federal contractors face this problem and failed in their attempts to establish a strategy that would work in the commercial arena. They could not respond quickly to the requirements in the commercial sector and were unable to expand their business beyond federal.

One other member of the management team with whom I had a close relationship with was Guy Barudin, the CFO. He was a highly experienced financial officer who had worked for a few Fortune 500 companies. One day, during a conversation with Guy, he mentioned to me a problem that was festering in the company. An outside advisor to Huggins convinced the management team that they could make higher margins than federal IT, which was generally 3 to 7 percent of revenue by diversifying their business into the construction area. This new business activity was initiated prior to me joining the company.

Guy mentioned that the company had bid on 15 federal construction contracts and won 14. I said that was great. Then Guy dropped a boom and said: "We underbid all the contracts we won due to the company's lack of internal experience in the construction business". I asked him about the decision-making process that was deployed before getting into this business. Guy stated that they retained a consultant who talked a good game, but he had no construction operating or business experience. Also, there was no one in the company monitoring the performance of the work and the workers were stealing the supplies and equipment when they were not getting paid. So, I suggested to Guy, "Why doesn't RMS just sell or close this entity?" He said, "We can't because the entity is RMS. The company never set up a separate company for the construction business." Wow, this was a major problem. My next suggestion was at a closed senior management meeting with Rotan and Guy; I suggested that the company draw down some of their bank line facilities or borrow some funding to resolve the suits and get out of these contracts. Guy mentioned that several of the federal construction entities were already taking action to have the government suspend all the company's other federal contracts. Guy informed me that next week, the RMS contracts with the federal government, FAA, and others were all going to be suspended and the bank account was frozen. The company was out of business and was under federal investigation. This $100 million company with more than 1,000 employees after 25 years was nearing the end of their remarkable track record in the federal contracting arena. The lesson learned is not to enter a market where you have no internal experience and set up a separate entity to protect your primary business.

David Huggins was a well-respected businessman in the New Jersey and Philadelphia community for more than 25 years, and people started to hear of his problem. Jerry Johnson, executive president at Safeguard Scientific spoke to Pete Musser, the president to see how they could help Huggins. Safeguard worked out a plan to restructure the company and take over its assets and liabilities. I am sad to say that David Huggins died within six months of Safeguard taking control of the company.

Following the change in ownership of RMS, I became a senior consultant at Safeguard Scientific during the dot-com boom/bubble period.

Chapter 11: The Dot-Com Boom
Network Solutions Value $21 Billion

Many people reading this book may be too young or were not yet born to recognize these words: the dot-com bubble. Even some older individuals who lost money during the bubble do not want to remember this period in our financial history. This was a critical period related to what happened in the history of Network Solutions.

In the article posted on the Business Insider website, "An economic bubble exists whenever the price of an asset that may be freely exchanged in a well-established market first soars then plummets over a sustained period of time at rates that are decoupled from the rate of growth of the income than might reasonably be expected to realize from owning or holding the asset."

Whenever this mismatch takes place the result is a financial bubble. Do you remember the housing bubble of 2008 when millions of people lost their homes and savings because of the decline of housing assets? Overzealous mortgage bankers provided loans for homes that were financed beyond their current and future value.

"Not since the South Sea Island bubble in the 1700s have Western economies experienced anything like the dot-com economic bubble. Suddenly everyone wanted a piece of the Internet action; normally astute investors went crazy and moms and dads added to the frenzy. For some, the dot-com era saw an amassing of great wealth. But almost overnight it disappeared from 2000 to 2003," by Ian Peter of NetHistory.

The dot-com bubble started in April 1997 through June 2003, five years of financial growth in the stock market and then a rapid decline in 2001 to 2003. During this period the Nasdaq peaked over 5000 on March 10, 2000, after doubling within the prior 12 months. The market had not experienced that type of growth in such a short span of time. Once the public got hooked on the Internet explosion, they were buying stocks of anything related to the Internet not wanting to miss the boom. The boom period commenced a little over two years following the sale of Network Solutions to SAIC.

During this period, I was employed as a senior consultant with Safeguard Scientific. Safeguard was headed by the late Pete Musser, a financial legend in the surrounding Philadelphia community and a very generous man. He made donations to numerous foundations and organizations. My working at Safeguard was ironic since it was the late Pete Musser who had turned down making an investment in Network Solutions in 1993 and, now, I was back at the building following the sale of Network Solutions. I was hired by Jerry Johnson, executive vice president and maybe the only African American to hold such a high position in a publicly owned venture capital firm at that time. This is still very rare today with few African Americans employed in the venture capital community even after 25 years.

Since 1955, Pete Musser had been successful with his investment company because he invested in companies that had a history of making money. For decades, he invested in stable companies that demonstrated the ability to grow and deliver a return to its investors. He was an early investor in Novell and convinced the company to move from hardware into software. Musser was one of the first to formalize the now popular incubation of startups, housing companies on the Safeguard campus. Many people acquired Safeguard stock because they believed in Pete and his conservative investment strategy. The other side of Pete's personality is his ability to recognize opportunities regardless of the ethnicity or gender of the person. Besides having an African American, Jerry Johnson, on his senior management team, Pete Musser provided loans or investments to a number of African American entrepreneurs; Cecil Barker Founder and President of OAO, was assisted in spinning off one division of his federal company entity into a commercial businesses that eventually became a public company; Willie Johnson, president of PRWT in Philadelphia provided funding for expansion; Della Clark, an African American woman in Philadelphia helped her grow the Enterprise Center. He also provided a personal commitment of $10 million for a proposed minority-owned venture fund that I personally developed. Additionally, he rescued a failing African American technology company, RMS from bankruptcy. The only mistakes Pete made were not investing in the African American controlled Network Solutions in 1993 when we offered him an equity share of the company. The second being, when the Internet boom started, Pete was persuaded by his young management team that the traditional investment model had changed in the market with the advent of the Internet. They convinced Pete to look to the future and start investing in companies with unproven growth potential by capitalizing from the explosive Internet and disregarding his long-established belief in only investing in companies with a profit or that are close to being profitable. In fact, Safeguard Scientific's stock price during this period rose by more than 300 percent due to its public promotion of investing in Internet technology companies. This enthusiasm to invest in Internet technology enterprises was not just at Safeguard, but on Wall Street and Silicon Valley, where most venture capital organizations were based. Many of the investments were in companies led by inexperienced and young entrepreneurs with no clue on how to launch a new marketing idea, retail or service concept, but only their belief in the "mystique" of the Internet: creating platforms to link associations, sports teams, developing new directories, marriage counseling, dating services, travel services, and more. At Safeguard's office, we saw hundreds of investment proposals each week and all the conference rooms were occupied with managers reviewing PowerPoint presentations on the next big dot-com idea. New companies were being created by the minute with investment presentations sometimes made on a napkin during lunch with the investment check to follow within 36 hours. These newly formed companies were now receiving millions of dollars to take advantage of the "ramp" that the original Network Solutions had created.

It had been less than three years since the partners at Network Solutions had sold their company for less than $5 million for the lack of capital. Now, the technology they helped establish was making hundreds of millions for other people and other companies.

This frenzy of investments in companies that didn't have any earnings continued for a couple of years. Since my career started as a banker with two noted financial institutions, I was trained not to make loans or investments in entities without a history of earnings. At one of my bank training programs at JPMorgan, I was taught by the late Sandy Burton, a CPA and renowned for his teaching students the tricks companies will play with earnings recognition in their financial statements. Also, how they could make adjustments in individual accounts to make their bottom line look better than it really was. This is one lesson I never forgot and have always evaluated financial statements with Sandy Burton's voice in my head: don't believe the numbers until you investigate how they were created.

In the case of the dot-com boom and the companies that were being created, many had no earnings and may never have any earnings, basing their investment strategy on utilizing this revolutionary technology, the Internet.

One of the people I stayed in close contact with at Safeguard was the chief technologist. Whenever we would meet, I would ask him, "When are the earnings coming?" and other research to justify such high stock prices for these new companies. Many of the investors were hoping to capitalize on the market frenzy around Internet companies. Most disregarded the negative factors of inexperienced management teams and no earning history. This would allow investors to get a high return on their investments even before the company showed any earnings based on hype. This strategy went on for several years while investors waited for revenue and earnings to show up. Safeguard had invested in a number of these companies and received a return from the individual company's stock and rocket-ship rise of Safeguard's stock. Every time Pete Musser would give a presentation to analysts, he would state that he was confident that earnings would come soon, so he never sold any of his personal stock during the dot-com boom and even bought more on margin. This strategy would later cost him dearly.

Another day when the market continued to climb, I cornered Safeguard's chief technologist again in the hallway and asked about his belief in the market and he pulled me aside. He stated, "Safeguard is changing their investment strategy after three years and were now focusing on investments in infrastructure". When the market got wind of Safeguard's new investment strategy, the *Wall Street Journal* published the story of the company's investment, now in infrastructure. All hell broke out on the financial markets. Safeguard was considered one of the "bellwethers" in the future direction of the Internet and everyone tracked their stock closely and their investments in these Internet companies.

Some of my friends that knew I worked for Safeguard called me. "Al, should we sell or stay in the market?" I would recommend most of them to get out and take their profits if they had any.

This is a situation where the saying, "If you build it, they will come" was wrong. They did not come for numerous reasons related to lack of infrastructure, the market still not being fully developed, and skepticism of the general public about the Internet's potential. People became oblivious to these combined problems after a couple of years of tremendous stock growth. This was all based on the Internet hype. Some of the major tech investors lost millions during this period. Even Jeff Bezos got caught up in the early hype and many of his investments failed. This did not stop him because he believed that the Internet was still a viable platform worth his investment of time and his investors' money.

The market immediately dropped by more than 500 points, bringing down many of these newly minted public dot-com companies, substantially reducing the price of Safeguard's stock portfolio. Many of the Safeguard managers had been selling shares of these companies' stock once they met the legal hold period and did not get hurt as much. However, Pete Musser believing that these companies would eventually make a profit lost everything and financial institutions said he needed to cover his margin account position (leverage loans); that he was unable to do. A *Wall Street Journal* article stated that Pete Musser had lost $1 billion during the dotcom bust. He was later able to reduce his losses with the help of friends and investors. He swears to never ever buy anything on margin again.

During the DOT BOOM, SAIC and NSI and other shareholders sold the balance of their shares to VeriSign, a California based technology company for $21 billion. The historical date was March 10, 2000. There were articles in all the major newspapers and on all the financial

TV networks discussing this financial sale of Network Solutions. *It was the largest sale of a technology (non-communications) company ever.* This was the second time that Network Solutions made financial history since the company became a public company.

The acquisition of Network Solutions by SAIC generated more than $3.2 billion in cash as stated in their book and assisted SAIC to financially restructure their company and created their own Venture Fund to fund other technology companies.

What is important about the dot-com boom was the escalation of the market value of Network Solutions during this period. With the frenzy of new Internet companies being created with no revenue or earnings, projected market values worth hundreds of millions of dollars based on a technology that an African American company controlled and provided access to the global community. But when the original owners tried to find investment dollars to maintain control of the Internet, none was available.

Network Solutions Historical Timeline

1979
Four founders incorporate
Network Solutions

1980
SBA certifies the company as an
8(A) minority owned firm

1985
Network Solutions President
Changed from Ty Grigsby to
Emmit McHenry

1990
Network Solution awarded the
highly publicized contract from
AT&T under the $50 Billion
Contract, FTS 2000, managing
the voice, data and video
throughout the United States

1991
Network Solutions appointed
subcontractor for the creation of
the Defense Department
Networks addressing system

1992
Network Solutions wins the
contract from National Science
Foundation to create the
platform to issue Internet
addresses globally for the first
time

1993
Network Solutions commenced
issuing Internet addresses to the
global community

1995
Network Solutions founders sell
company to SAIC

National Science Foundation
approved the charging of a fee
six months following the sale of
Network Solutions.

1997
Network Solutions becomes a
publicly traded company valued
at $382 Million

1999
Network Solutions under new
ownership sells stock in a
secondary offering, company
valued at $779 million, the
largest Internet Company in the
world based on market value

2000
Network Solutions is sold by
SAIC to Verisign for $21 Billion,
historically the largest sale of a
technology company at that time
outside of the telecom industry.
In todays dollars the value of the
company would be worth $40
Billion.

Ownership of Network Solutions Following the SAIC Sale

Network Solutions has been acquired five times since VeriSign acquired the company in 2000 for $21 billion. Below is a list of various owners of Network Solutions and the prices they paid for the company.

On October 17, 2003, VeriSign announced the sale of the registrar business, which continued to operate under the branding of Network Solutions, to Pivotal Equity Group for $20 million (VeriSign still retains the registry business which had been originally created within Network Solutions prior to VeriSign's acquisition of the company.)

During January 2006, Network Solutions acquired the company Monster Commerce, cofounded by Stephanie Leffler and Ryan Noble in Belleville, Illinois.

On February 6, 2007, Network Solutions announced that General Atlantic, a private equity firm, would acquire Network Solutions from Najafi Companies (formerly Pivotal Private Equity). Although terms of the deal were not released, the *Wall Street Journal* reported in a story on May 30, 2007, that the price tag was "around $800 million."

At the end of July 2007, Network Solutions had 6,659,150 domains under management and was in the top five wholesale domain registrars following GoDaddy with 19,709,215 domains and eNom with 7,646,676 domains.

In August 2011, it was announced that Network Solutions would be bought by Web.com for $405 million and 18 million shares. On October 27, 2011, Web.com announced the completion of the acquisition.

Network Solutions today is a subsidiary of Web.com, a company located in Jacksonville, Florida. Siris Capital acquired Web.com for

$2.2 billion in cash, October of 2018. Siris is in New York and Frank Baker Managing Partner and Co-Founder is a highly successful investor in the tech space.

When Network Solutions Accidentally Shut down the Global Internet

Following the sale of the original Network Solutions, a historical event occurred signifying the power of Network Solutions with global control of the Internet.

The events of July 17, 1997 validate this point. Two years following the sale of Network Solutions to SAIC, the global Internet totally stopped working. Just think of that happening today, with no Facebook, Instagram, no streaming of videos or music. How many people would go crazy until the Internet was operating again? Today, more than 4.2 billion people have access to the Internet, half the world's population.

In the book, *Names, Numbers, and Network Solutions: The Monetization of the Internet* written by J. Robert Beyster and Michael Daniels, they specifically state that Network Solutions' error in loading data files of Internet addresses, on July 17, 1997, stopped the Internet. The Clinton Administration and governments around the world pointed the finger at Network Solutions, the manager of the Internet address files, for this disaster and the disruption of the Internet for more than a few days. It was apparent from this incident that Network Solutions' role as providing Internet addresses to the world also meant they could basically stop the operation of the Internet, meaning they controlled the Internet. There are numerous other organizations in the communications industry that provide the cables and wires supporting the operation of the Internet as well as networks around the world that link to the Internet. However, it was Network Solutions' activities in managing the address system and providing this information to the world that is critical to the operation of the global Internet. The human error that occurred of the Network Solutions' employee to shut the Internet down by not providing complete files and records for Internet users around the world. When these files did not have the critical information related to the Internet addresses, the Internet had no purpose in operating, resulting in the system shutting down all over the world. Think of Network Solutions as the source of blood to the human body. Without blood in our system, the human body stops functioning and all our organs shut down. That's what happened in 1997, the system shut down around the world because someone forgot to load the file of current Internet addresses.

This could not happen today since Network Solutions is not the only provider of Internet addresses. It would be a massive error in all the other providers of URL addresses to shut the global system down again.

Part II: Providing Solutions and Direction for the African American Community

Chapter 12: The Current Wealth Gap of the African American Community

One of the major reasons that this section of the book highlights some economic statistics concerning our community today brings attention to what we need to do. These issues impact our community's ability to provide economic security for our children and grandchildren in the decades to come.

During my youth, I often heard statements made by my parents and elders in our community who wanted to ensure a better life for their children than they experienced themselves. This entailed providing a good education so that they could earn a higher standard of living and acquire assets (houses and other property) to enhance their ability to live a comfortable life with little or no debt. When my parents passed, they left their children (me and my brother Martin) assets that could be shared with our children. These assets were in the form of money to pay for our children's college educations, to pay down loans, and for down payments on property. The experience of my parents growing up in the segregated South and moving to the North in the '30s, like many other African Americans during the Great Migration, paved the way for this generational focus of providing a foundation for future generations. During the '60s, '70s, and '80s we made some gains in our economic wealth in order to provide financial support for future generations. Recent statistics demonstrate that our community is generating and will generate very little economic wealth for future generations unless we do something about the problem now. We are at a point in the history of African Americans in our country that the economic growth of our communities has substantially declined. The numbers are astounding, the level of wealth (net worth of the black family) today is 10 percent ($17,000) of the wealth of the white community ($171,000). Even the Hispanic community wealth is higher than that of the African American community at $20,720 is shown below from the Federal Reserve Board study in 2017!

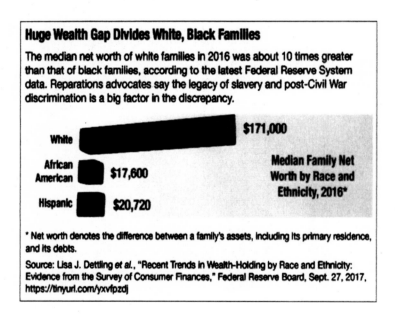

Huge Wealth Gap Divides White, Black Families

The median net worth of white families in 2016 was about 10 times greater than that of black families, according to the latest Federal Reserve System data. Reparations advocates say the legacy of slavery and post-Civil War discrimination is a big factor in the discrepancy.

White — $171,000

African American — $17,600

Hispanic — $20,720

Median Family Net Worth by Race and Ethnicity, 2016*

* Net worth denotes the difference between a family's assets, including its primary residence, and its debts.

Source: Lisa J. Dettling et al., "Recent Trends in Wealth-Holding by Race and Ethnicity: Evidence from the Survey of Consumer Finances," Federal Reserve Board, Sept. 27, 2017, https://tinyurl.com/yxvfpzdj

The recession of 2008, with its enormous decline in property values, hit our community harder than any other community in the U.S. In fact, we lost more than 30 percent of our economic wealth since our homes are our major assets compared to the white population which had other income sources from ownership of securities, savings in the bank, and ownership of rental properties.

A 2019 study conducted on the economic impact of closing the racial wealth gap by McKinsey & Company, stated that the widening of the racial wealth gap of the disadvantaged black families, individuals, and communities limits economic power and prosperity in our country. From McKinsey's research, they believe that 70 percent of middle-class black children are likely to fall out of the middle class as adults.

We have several problems that we can fix; if we work together, trust each other, and believe and support the ability of individuals in our community. This is the main reason for my addressing these issues related to a better understanding of technology opportunities today, in order to close the wealth gap for black people. The white community contributes 34 percent of their income from equity ownership, mainly business ownership. We contributed less than 14 percent, reducing our economic wealth. The main issue facing our community today is our high level of debt compared to other communities. This is sapping our families and young people of their potential. With the real possibility of a recession in 2020 or 2021, the impact on our community could be devastating unless we take some action mentioned later in my book. Our greatest threat is our inability to understand and predict the changes that will take place in our economy related to future technology. Also, we need a better appreciation and to understand future opportunities in various economic arenas: real estate ownership, technology, and education, all resulting in wealth building. It is good to be mindful of our current situation, but economic empowerment entails knowing what the future can provide and ensuring that our young people are trained and educated so that the next recession does not result in a depression for us.

The African American Community has faced numerous obstacles for more than four hundred years and still have been able to achieve great things. However, our forefathers understood that they could not solve any economic, political and social issues if they did not work together and provide a foundation for our children and the next generations. We are in a race to rebuild our economic wealth. It is no longer just a local race for opportunities in the U.S. but a global competition from outside our borders. Those immigrants trying to build wealth in the U.S. have a vision of future opportunities for their family and community. We need to return to this type of thinking where we are working together and believing in each other's abilities.

We can change these numbers since we have the purchasing power and political strength to influence our environment and increase our per capita income over the next decade. In the next couple of chapters, I provide a "road map" of potential business in the technology and non-technology areas to grow job opportunities in the second Internet revolution.

"Change can be frightening, and the temptation is often to resist it. But change almost always provides opportunities— to learn new things, to rethink tired processes, and to improve the way we work."

– Klaus Schwab

Chapter 13: Business and Job Opportunities in the Internet Economy

In the last two decades, we have seen thousands of entities that have taken advantage of the Internet's global communications network. This technology has really changed since my presentations to audiences across the country on the potential of the Internet in the '90s. I did not consider that there would be an Amazon or Alibaba, to provide a global platform to sell and distribute products anywhere in the world. Amazon started with selling books, but now you can buy virtually everything you want and have delivered to your home or office.

The beauty of Amazon for businesspeople today is they don't have to worry about the distribution of their products which was a major issue before the Internet. Warehousing and delivery services were major business obstacles before Amazon. Also, providing a platform to advertise and market your products has been solved for these Internet-based companies.

So, what today's entrepreneurs need is a marketable product or service. Then, they can turn over the rest to Amazon or other global marketing companies. In fact, I have a contract with Amazon to sell my book. The self-publishing industry has become huge due to the ability to instantly have your product in the market after the writing of your book is completed. Additionally, Amazon even provides a rating of your book from their sales information compared to other books that they sell in your area of focus. For these general sales services, Amazon does charge a healthy fee but considering all the services that you receive that allows you to focus on your product and have an instant market overnight, it is reasonable.

A couple of years ago, I met a young couple who were seeking to commercialize anti-aging and cosmetic lines they were planning to sell on Amazon. They told me this incredible story of how they became Amazon's leading product in this category overnight. They were generating more than $12 million in revenue in two years with six employees. They were looking to double their sales in the next year.

This company got started after obtaining research information from Google, on the most frequently asked questions about women's healthcare and cosmetic issues. They tracked this information for about six to eight months and then started looking for cosmetic and anti-aging products that could solve many of these issues. One of the partners, a cosmetologist, was familiar with some of the issues and knew of certain products that were not widely distributed. So, these two entrepreneurs created a line of products and established a website and a distribution contract with Amazon to sell their product. They also had to contract with a couple of manufacturers to produce their formulation and to package the product. Today, this company is doing more than $50 million after only four years and with only approximately 10 employees. They told me that their profit or net income was much greater than they had ever expected. They were receiving calls daily from potential buyers of their company at a price of several times their annual revenue. The success of two well-known individuals making millions in the $8 billion cosmetic industry, Rihanna and Kylie Jenner, combines an online strategy with retailing of their products. Ms. Jenner just sold half of her company for half a billion dollars.

There are companies with similar experiences being created every day with a great product and the motivation to build a business. As with the successful cosmetic and anti-aging companies, they started with extensive research of the market and created their product based on the demand statistics. Before the Internet, this would have been time-consuming and sometimes too costly. Today with the online research tools available, these tasks can be done in a couple of hours, days, or months and, literally, at no cost.

The Internet created a global market for any product, and with social media sites, you can direct your marketing campaign to specific markets around the world. Amazon has access to billions of people. Recently, Amazon announced they have more than three million small businesses around the world taking advantage of their distribution platform.

I am sure there are buyers for your product or service. I failed to mention that general services are a growing market on the Internet. Need a plumber or someone to fix your car within 10 miles of your residence? If you can edit, provide legal services, translate, or just provide advice on how to fix something, it's all being marketed on the Internet. Location is not a problem since the Internet will allow you to receive a service in your home or office and receive payment through PayPal or credit card accounts. Airbnb has created a multibillion-dollar business renting rooms and housing units all over the world, using an e-mail address and Internet connection. There are numerous companies today that can help you set up an Internet platform to sell products and services. They can even arrange for you to source your product abroad or domestically. The entities providing this type of guidance have been able to assist motivated businesspeople in becoming millionaires with their marketing formula.

There are many energetic and motivated individuals who have created their own podcasts or YouTube videos that have led to a film or reality shows on cable networks. Many of them have signed long-term contracts. It just started with an idea and motivation. This is also happening in the music industry: broadcasting a new artist on Facebook, a podcast, or YouTube and creating an opportunity for someone who is trying to showcase their talents. You can create your own talent agency on the Internet and introduce individuals and groups to the general public. Remember there are billions of people around the world linked to the Internet.

Due to my association with Dr. Lonnie Johnson, the inventor of recreational toys, I had the opportunity to meet with a major toy executive and discuss the industry. One of the amazing things that is happening in the development of new toys by kids. This is just the reverse of the traditional model of toy companies identifying a future toy idea. With the use of the Internet, young children have created YouTube videos to share toy ideas with other children and get their feedback and suggestions. These toy executives stated to me that they are regularly monitoring these sites for the next toy idea. Additionally, there are children creating their own online marketing with their peers. There is nine-year-old, Giana who has over 22,800 Instagram followers. She acts as an influencer (a person that people follow) for fashion around the world in her age group. The *Wall Street Journal* did a feature article on Giana, a member of Generation Z, with roughly 67 million individuals born between 1997 and a few years ago. They have the purchasing power of about $44 billion with the assistance of their parents. The *Wall Street Journal* quotes, "Thanks to social media, members of the Gen Z see a staggering array of merchandise and pinpoint precisely the clothes and shoes they want to wear. Almost half are a race other than non-Hispanic white." As in the toy industry, these young people are changing the way marketing is done today and, in the future.

One innovative and imaginative African American woman in Cincinnati, Ohio, Bethany Gaskin, started a YouTube channel in 2017 showing her eating seafood. She would place a full table of various types of seafood like king crab legs, mussels, and lobster tails sometimes with hard-boiled eggs and roasted red potatoes. Would you believe that hundreds of thousands of people tune in each week to watch Bethany Gaskin binge eat shellfish on YouTube? Gaskin has capitalized on the popularity of a food video genre known as "Mukbang," states *New York Times* writer Jasmine Barmore.

Mukbang began as an Internet trend more than a decade ago in South Korea. The videos produced in this unemployed housewife's home have made her a millionaire, allowing her to employ her husband and other family members. Most viewers enjoy seeing Gaskin eating these seafood items. The writer states that viewers who have eating disorders or are stressed out take pleasure in seeing someone who really enjoys eating. The two YouTube channels, Bloveslife and BlovesASMR, have made millions for this wife from Cincinnati, Ohio with 1.8 million YouTube subscribers, and a following of nearly 900,000 on Instagram. She is making money from advertisers and fees from the providers that pay her for access to her audience base and sell their own products or services.

Finally, something that we give very little thought to, but is a booming business, is the pet industry, with more than $70 billion spent annually. The dollars spent in the pet industry annually exceed the annual dollars spent on babies. This is a growing industry with everything from pet hotels to grooming services that are being opened every day. It was recently announced that Softbank, a leading venture capital company, invested $460 million to support an Uber-like car service for pet owners.

If you are a pet owner, you may have an idea that can improve the life of a pet or its owners. You may want to conduct a Google search to see if anyone else has already thought of your ideas.

I am providing these examples in order to stimulate you to think about which Internet product or services you can monetize. You still have plenty of time to make your millions on the Internet like Bethany Gaskin.

Non-Technology Business Opportunities for Today

There are several non-technology-based products and services that have a high demand and will for many years to come. One business that I believe will continue to grow is the disaster service area. I am sorry to say this, but it is a major growth area in our economy today and in the future. In fact, the number of disasters in our country is growing, from hurricanes and floods to out-of-control fires. This is further impacted by our inability to get climate change under control with some people believing that this is not the reason for these disasters.

The official figures are in: 2017 was among the three warmest years on record, according to an announcement from NASA and the National Oceanic and Atmospheric Administration (NOAA). "Last year may go down in history as the year that the impacts of climate change finally became undeniable," states Chris Webb of Worldwide Fund for Nature (WWF), the world's leading wildlife conservation organization. While writing this book, we had major floods along the Mississippi River and wildfires in California that destroyed homes and cost billions of dollars. We are still trying to recover from Hurricane Harvey and other disasters in Puerto Rico, Florida, and Texas.

The dollars spent on these disasters in Texas, Florida, Puerto Rico, and the Virgin Islands have cost U.S. taxpayers around $350 billion in disaster relief and rebuilding of these areas. Since the Katrina disaster in 2005 in Louisiana and Mississippi, the dollars required for disaster services and products have increased by almost 150 percent in little less than 15 years.

There are opportunities for commercial inspection companies, environmental detection services, building materials, construction contractors, relief workers, documentation specialists, insurance adjusters, trucking and hauling, fire retardant providers, mold and mildew detection, and remediation companies. These are just a few of the opportunities in this growth industry. In many cases, the nature of the work continues for five to seven years following the disaster. Then follow-up services continue for 10 to 12 years, related to administrative reports. In fact, contractors are still providing administrative services to state authorities for the Katrina disaster which occurred in 2005. U.S. Department of Housing and Urban Development (HUD), the major provider of funds to rebuild, requires the tracking of records for almost 10 years following a disaster where Community Development Block Grant (CDBG) funds are utilized. Yes, there are several major players in this area, but they are all required to meet federal regulations related to contracting with small and minority businesses. Additionally, the state and local governments must meet small and minority business contract regulations since they are utilizing mainly federal funds to rebuild.

The infrastructure opportunities will be enormous over the next decade. There are more than $4.6 trillion in projected infrastructure repairs required in our country during this period: bridges, highways, schools, and federal and state structures. Most airports have expansion programs in place to handle the increasing number of travelers and the new services being offered at airports. Airport concessions are a booming business today and there are many African American businesspeople taking advantage of this growth area.

Another major opportunity generating more than $500 million in grants and loans from Congress, the U.S. Department of Agriculture, the National Telecommunications and Information Administration (NTIA) is broadband technology expansion. This is a long-overlooked problem since the launch of the Internet in 1993, labeled the digital divide problem. The Federal Communications Commission (FCC) states that more than 21 million people lack fixed broadband to enable their communities today to get up-to-date healthcare information, access to telemedicine services, latest educational studies, agriculture information, and disaster communications. Many of these communities are in the southern region of the country, but cities like Detroit and Memphis have large areas without updated broadband access. There is a need for businesses to build out these communities with broadband telecom products and services. This is an area where African American businesses need to develop joint ventures or partnerships to obtain grants and loans for the buildout of these communities. We have a large pool of experienced telecom engineers and technicians who can win contracts to help these communities gain access to the latest broadband technology. Many of the communities affected are African American. These areas of the country will be overlooked by industries without a labor force with the necessary training and skills delivered by broadband technology. Broadband is an enabler technology that should be mandated for every community in our country like many countries in Europe and Asia.

If you are seeking funding for your inventions or applications in health, military, cybersecurity, transportation, and energy, federal agencies are waiting to give you money. In fact, the federal government assists inventors with funding to develop your new products that have applications suitable for federal and commercial customers. The federal government will provide inventors with grants to help launch their products. The Small Business Innovative Research (SBIR) program. This program is supported by most federal agencies and grants can range from $150,000 to a couple of million dollars. Additionally, if you are looking for a product, there are 15 federal research labs with extensive patents and technology portfolios that are available to the public for commercial development. Many of the technologies we see in cars today like sensors and directional information (GPS) were developed by the federal government and later licensed to private companies to take advantage of opportunities in the commercial sector.

If you wait too long, someone else will come up with a similar idea and you will say" I had that idea several months ago". If it does not work, then look for something else, but do something. Another area of great growth is the healthcare sector.

With more than 70 million baby boomers seeking assistance in their everyday living, there are business opportunities in providing some type of service or product. In the healthcare area, the number of visits people are making to hospitals and the amount of time spent in these facilities have become very costly. Federal, state and local healthcare professionals are seeking ways to monitor these patients' activities and vital signs remotely. This has developed into an enormous growth opportunity (trillion-dollar industry) for home care services (caretakers or nursing assistants). As a greater number of people live longer, this type of service will continue to grow with expanded services that include prepackaged food delivery as part of the home care service. It would not be surprising to see Amazon enter this business since they already own Whole Foods and have an established delivery system. Now, this could be enhanced with their recent purchase of the Ring Monitoring System to track what is happening in the home from the caretaker to what's in the refrigerator through artificial intelligence which I will discuss fully later.

You need to think out of the "box" for new and imaginative business ideas. Make something happen.

Chapter 14: The Internet Today and Its Direction for the Future

I am a strong proponent of knowing what is going to happen in the future with technology and other research areas. Understanding future trends help you cope with unexpected changes in the economy which could jeopardize your employment situation and the well-being of your family or community. In this section, I want to point out some trends that you need to understand today.

In the '90s, when individuals were test driving or testing out their Internet addresses, people were able to access electronic messages and information that they never had seen before. The performance and capability gradually improved with high-speed networks and the ability to send and receive data files allowed pictures and video to be viewed easily.

I remember when my youngest daughter, Adrian, mentioned in 2004 that her college, Hampton University, had been introduced to this service called Facebook where students could share their profiles and make contact with other students over the Internet. It was first started at Harvard by Mark Zuckerberg. This technology started to grow on college campuses and then was introduced to the world. Today, Facebook has more than 1.9 billion users every day, and revenue of $16.9 billion with profits of $6.9 billion. What we did not expect from the Facebook technology was a tool being deployed by foreign governments and individuals to tamper with our elections.

Technology solutions have been developed over the last 25 years utilizing the Internet platform to change our everyday life. The influence of Facebook and other Internet-related businesses will continue to grow over the years.

There has been greater advancement in technology over the past 25 years. Thomas Freidman stated in his book, *Thank You for Being Late: An Optimist's Guide for Thriving in the Age of Accelerations,* that most of the Internet-related technology companies we see today were already established in 2007 or were formally launched: Uber, Lift, Instagram, YouTube, and Airbnb.

Once technology companies realized the breadth of what could be done to bring people together on and with Internet technology, knowing an individual's likes and dislikes, radically changed society. It has led to providing services based on access to information quickly, to pick you up and find a room or house in any city, or car services like Uber, Lyft and Airbnb, what is next?

The Internet has reduced the time it takes to bring ideas to the market and place them in the hands of consumers. People today have access in minutes to vast sources of market research, reports, and customer profiles. Most of the new technology tracks and monitors your response to questions and buying habits through your smartphone, tablet, or personal computer.

I am sure you have noticed that your every movement can be tracked by the device you are carrying. If you are in a restaurant or store, your cell phone can identify your location and the name of the store where you are shopping. You can also be sent notices of sales for the particular store or suggestions of what to order at the restaurant where you are.

The technology that you are hearing more about each day is artificial intelligence or AI. Ted Greenwald of the *Wall Street Journal* states "AI is anything a computer can do formerly that was considered a job for humans." Anything with the word "intelligence" makes me think of information that the FBI or CIA uses. I am not far off; this information is begin collected from devices that you use every day. This information is stored in a computer in the "cloud." Yes, that is another scary word that does not represent the physical cloud we see in the sky, but storage facilities that hold enormous amounts of information until it is needed. AI technology has your personal information in the cloud to call upon to market, sell, or direct personalized information. Another term linked to AI is machine learning, a process of taking AI and massaging this information with statistical analysis of real-world phenomena to make predictions. These systems required a huge amount of data to provide the efficiency of the process that was provided by the advent of the Internet. With enormous amounts of data generated by Google, Amazon, Facebook, and Microsoft with cloud storage technology, AI and machine learning blew up.

Who is collecting this information and what do they plan to do with it? With the advancement in computer technology, information can be processed regardless of the size of the file in nanoseconds. The information being processed can be correlated with other information related to your profile. With this increase in computer speed, your genetic profile and personal attributes are tracked every time you use a social media platform. In the past, this process was performed by humans, but today it is done through, AI. Decisions are being made by computers. This type of technology is being deployed in both public and non-public entities. In the future, through AI, decisions on your children's schools, your job, your future husband, or your wife will utilize this technology. Most dating websites today use some form of AI to link profiles to determine compatibility between people. However, there is still some level of human involvement in these decisions. So, the question again, as the power of computers continues to increase and AI crosses more information platforms, computers will be making more decisions for us in the future. You will hear about Artificial Intelligence more over the years and you should know how it will impact your life and that of your family as well as your community.

Sundar Pichai, CEO of Google in 2018 stated: "AI is one the most important technologies we are working on today. It is more profound than electricity and fire. We have learned to harness fire for the benefits of humanity, but we had to overcome its downsides too."

These uses of technology are playing a greater role in the decision-making processes, sometimes consciously or unconsciously, and eliminating wasted time. Self-driven vehicles, not only cars but trucks and buses, are being tested on our roadways today. In several cases, the rush to get this technology on the road has led to recent fatalities. Once the safety problems have been resolved, expect to see self-driven vehicles in your community by 2023. Many believe once autonomous vehicle technology is perfected that traffic fatalities will drop substantially.

The thing that makes all this possible is the development of sensors and devices that are being deployed in our homes and throughout society. Technology exists today to reduce the number of hours humans work and the type of work that they will do in the future. These changes will take place in half the time technology changes have taken in the past.

Sensor technology that I mentioned earlier is a new technology platform, the Internet of Things (IoT). This is an integration of technologies that are enhancing the capability of the Internet and providing new ways to monitor and track things in your house, car, office, health club, and even in managing your pets. IoT is a network of devices that are being introduced everywhere you are and everywhere you are going or thinking about going. For example, the driverless car: the foundation of driverless vehicles are sensors. These sensors determine where cars can and cannot drive, and at what speed. Sensors are communicating with other sensors within the vehicle, along the roadway, and with other vehicles to determine weather conditions, distances between vehicles, and obstructions. Late-model cars have sensors deployed to identify the car next to you or in front of you, how your car is driving, and any problem with your engine or lights. They are even developing sensors to track the mood of the driver as well as his or her driving capacity.

I recently did a radio show on Sirius XM radio called "Inside the Issues" with Wilmer Leon. I mentioned the driverless truck and a caller who I assume was a truck driver stated that this could never happen because of the challenges of driving a truck under difficult road conditions, such as on an icy road. The caller believed that a driverless truck would never be able to handle such a situation. I mentioned to him that the truck's computer would have instructed him or the onboard computer not to drive that day because of the forecasted weather conditions in the area 24 hours before his trip and let him know of changing weather and road conditions in minutes.

The Internet has enhanced computer and cell phone technology has even reduced the size of the workplace and likewise, substantially reduced the labor force of administrative and clerical personnel. The advancement in our telecommunications technology has made it easier to communicate with clients or staff without being in the same office or geographical area. In fact, you can see them while you are talking to them, through teleconferencing technology. If you do not care to have the other person see that you are at the beach or on the golf course, you can do a simple conference call on your cell phone with all the participants. Can you imagine the amount of wasted time and money that has been saved in traveling to another city, state, or country before conference calls and video call technology was developed?

This same type of technology advancement and reduction of travel time is being experienced with the advent of new Internet technology. The explosion taking place with rental bikes and electric scooters on just about every block in America and abroad was developed with the help of an Internet tracking and communications system. The Uber car service was born from the advancement in mapping and communications system that utilizes the Internet to provide your location and send a car to pick you up. Besides Uber-type car services creating the loss of revenue of traditional taxi drivers in New York and other cities, management of public transit systems are considering reducing their employee headcount if their revenue continues to decline as the expanded modes of these convenient travel options are more widely used. In less than 10 years, Uber and Lyft have already changed the face of basic transportation in virtually every city in the world. Medallion cabs in New York were valued at more than $1 million only several years ago. The price of the medallion cab in New York is now less than $150,000 and continues to drop. Uber and Lyft drivers have more flexibility to work based on their schedules and keep a higher percentage of the fares. These companies did not need to make a capital investment in vehicles but in tracking and monitoring software.

Now, technological advancements are reducing the number of jobs in shopping malls and supermarkets. Even McDonald's has instituted a computerized ordering and delivery system in most cities. There are more and more fast-food restaurants experimenting with automated grill robots to handle preparing your next hamburger and

French fries. The expected shortage of employees is driving the changes in these industries at breakneck speed. Every major restaurant in the world has mechanized their ordering and delivery system; if they do not, the out-of-business sign will be hanging on their door.

If you do not understand the changes that are taking place, you might become an unemployment statistic!

Most manufacturing facilities in the world already utilize some form of sensor or monitoring devices. In the past, people were deployed to physically check the quality of a product being manufactured. Even the movement of the product was done by conveyor belts that were controlled by humans. Over the years, sensing devices and robots have taken over these tasks and made the manufacturing floor more efficient and increased the level of productivity. Foreign and U.S. car manufacturers have been utilizing these types of technology platforms for years, substantially reducing the number of jobs in all types of factories throughout the world. The rapid increase in the use of Electric vehicles over the next decade will have a major impact on jobs in the auto industry. It is already estimated that jobs in the industry will be substantially decreased because electric vehicles require fewer parts and assembly, compared to gas and diesel cars. In fact, EV requires one third the number of workers to build an energy-efficient vehicle. The auto industry will change drastically in the next several years with increase robotics and reduction in the number of parts for Electric Vehicles.

This is the main reason why unions have been fighting to keep their members employed. They know this is an uphill battle, difficult for unions to win. The U.S. lost more than five million factory jobs since 2000 while manufacturing grew by 2.2 percent per year from

2006 to 2013. Contrary to the arguments put forth by many conservatives, American job loss has been more due to automation than to trade.

A study by Ball University stated that trade accounted for 13 percent of the loss of U.S. factory jobs, but 88 percent of the jobs were taken by robots and other types of mechanization.

We are now starting to enter the next phase called the second Internet revolution that will have an even greater impact on the lifestyle and the quality of your life than the first stage of the Internet. It has taken two decades for the Internet to be really appreciated, accepted, and widely used.

Impact of These Changes on Future Employment

This next phase is very important and should not be ignored. Technology development will change the way people live forever and their ability to provide economic resources for their families and communities in the future.

Let me give you a fact about the future; 40 percent to 50 percent of all jobs are in jeopardy of being eliminated within the next 10 years or so. The loss of jobs will depend on the nature of the work. But people at all economic levels should be concerned with this next phase of the technology revolution. This is not just being experienced in the U.S., but globally. In a recent report by McKinsey & Company, a nationally known research and consulting firm, "Our scenarios suggest that by 2030, 75 million to 375 million workers (3 to 14 percent of the global workforce) will switch occupational categories. Moreover, all workers will need to adapt, as their occupations evolve alongside increasingly capable machines."

The technologies I mentioned earlier are a major reason for these changes and will impact every sector of our economy to reduce headcount. If you do not have the right training and skills, you may not be able to find gainful employment in this mechanized economy.

Robots and mechanization will rise from the global average of around 10 percent across all manufacturing industries to around 25 percent by 2025. At the same time, IoT growth will be around $25 trillion and more than 50 billion devices will be in use.

Areas most threatened by automation are the operators of all types of machinery, back-office operations, banking functions, mortgage functionality, and any vehicle operation (cars, trucks, and buses). Collecting and processing data will ultimately be replaced by computers and robots. These devices will increase productivity because they are faster and do not need a coffee break or lunch.

The next automation revolution will go deeper and broader. It is likely that the management of investments and advisory services/traders will be automated within the next five to 10 years. There are a few hedge funds (money managers) that oversee management pension funds and other pools of funds, that are now testing automated trading services where a computer makes the selection for your portfolio rather than a human. This is being enhanced using AI.

If you are seeing automation taking place around you, then your job could be next, subject to a computer or robot replacing you. This is a really serious issue for today's millennials and their future. When the computer became widely used in the '60s and '70s, many people did not want to accept the technology nor the changes. Those who ignored the use and capability of computers were slow to adapt and were soon out of work.

With the combination of AI and IoT technology being deployed today, in the near term, we are going to face some real challenges related to employment opportunities. The real benefit of the Internet in the last 20 years has been the ability to provide information within minutes or seconds. Computer experts believe that computers and new Internet technological solutions are moving faster than humans will be able to adapt to these new changes. Am I scaring you? I hope so; this is a serious problem for all segments of our community.

These are some employment areas that will be protected from immediate mechanization and robots.

Healthcare (nurses, doctors, dentists, physical therapists, etc.): This is an area where human contact is very important to the execution of the job. The increase of an aging America will also require human contact that cannot be replaced by a robot. But there are several startup technology companies looking for ways to reduce the body count in this area in the near term. (There are systems being developed that will complement or enhance the worker's ability to do their jobs more efficiently and safely). You may not know this, but even the hospital operating rooms are already mechanized with robots to make critical incisions and removal of tissue. They will still need people to manage these devices and assist the patient in recovery.

Social Services: Social workers and counselors will have some protection from automation. However, decisions about job performance may be determined under the advice or guidance of a computer. Many family services will be analyzed by smart computers and AI techniques to determine the benefits and level of services from the correlation and integration of historical data. Social workers and counselors will need to know how to utilize information being correlated with historical data and diagnostic information, provided by computers. This area is being carefully watched by institutions and organizations concerned about AI technology making decisions that could eliminate opportunities for the least of us. Decisions made by computers with the assistance of AI will be based on data and no human input. For example, someone could press a button and access an employee record or health file and determine you are not eligible for financial assistance because of data correlation and not your human condition. What about a person not meeting standards determined by AI technology that results in eliminating you or your family member from the opportunity to attend college? These are very likely trends in the future as technology evolves.

Engineers and Cybersecurity experts: These are positions that will be necessary and will not be easily replaced by a robot. Someone must support the programming of the robotic machines. Writing codes and instructions to control the function of these mechanized devices will always be critical positions. Cybersecurity is a growth area and skilled individuals will be necessary to identify the various techniques of how hackers are penetrating networks. It takes a human mind to understand how another human attempts to steal or disrupt a network. Computers will be utilized to provide the analysis of human input for detecting some other human's devious strategy.

Executives and Managers: These are positions where human decisions are required in executing strategies determined by AI analysis. Humans are needed to make decisions regarding who provides value-added services and the ability to work within a team. However, there are other decisions that executives have made in the past, such as who should be promoted or relocated, that can now be decided by a computer. Voice-activation machines can even execute the hiring and firing functions. Managers will still be needed to determine the limits of technology and provide some analysis of potential issues related to these rapid changes.

Professionals and Inventors: Scientists and inventors are still needed to conceptualize and develop new ideas. The human mind is still important in pursuing solutions that will provide a cure for cancer or the next type of energy storage. Computers and robots will be involved in reducing the number of options and correlation of data and analysis. How do we solve the issue of global warming and create solutions to impede its damage to society for future generations?

Legal and Accounting professionals: If you are a member of these professional groups you need to go beyond your normal tasks more abreast of technology. Becoming more aware of how AI technology is changing the way decisions are being made in the public and private sectors of our economy.

Creative: This is a small group but is a growing category. Artists, performers, and entertainers will continue to flourish without the influence of computer technology or AI. Even though movies today and in the future utilize all sorts of robots, there will be a need for humans to provide the outcome and storyline that real humans can understand. We will always want to see humans provide musical talent, dancing, acting, and other creative activities.

Educational Professionals: Teaching will change over the next few years with robots in the classroom providing information and answering questions from students in areas too complex for unskilled and untrained teachers. Teachers will need to quickly advance with technology if they want to keep their jobs. With the new technology, students are demanding more information and instantly. They will not wait for the answer that can only be solved by mechanized devices that process large databases of information. The fact that today we are asking questions of a computerized box in our home, car, or office will change our whole educational system. We will start to see more students using their communications devices to get answers to questions rather than researching the answers. What will our brains be used for in the future?

The rise of video over the next few years will have a major impact on the learning process. This transition to a video-based society could substantially reduce the number of institutions, colleges, and universities. With the high cost of education and the enormous debt load that students and families are bearing today, video education in every classroom and home will be strongly seen as a solution.

Mary Meeker, a noted Internet expert states, "Video constituted 64 percent of Internet traffic and 55 percent of mobile traffic in 2014." Cisco, a leader in server technology, thinks that video deployment five years from now will include more than 80 percent of the entire Internet being online video technology.

These changes are just around the corner (some are already here) and we have not yet found solutions to this major problem of displacement of humans. We do not have the same lead time that existed for past technology shifts. Instead of 50 years or even 20 years, it will more likely take five years for future structural changes in our workplaces and likewise, our society. Our population is growing at less than 1.5 percent a year, so we must take into consideration this level of growth and the jobs that will be displaced due to the advancement in IoT and AI applications. What can we do today to stave off this future problem?

There are more than six million job openings today (2020) that can't be filled because of the lack of trained or skilled people. In U.S. Industry that understands these problems, more and more are seeking technology to solve this problem and fill this void with other than humans.

However, in the future, many of these job openings will require a higher level of technical skills and training. There are shortages of skilled computer specialists with AI and IoT experience. By 2030, the number of job openings will double and create a boom for coders, software application developers, system software analysts, computer user support specialists, and computer programmers. There will be more than a trillion computers in operation in 10 years that will require human direction and supervision. You need to have the math and science skills to obtain these positions.

Another technology that is fast becoming accepted and is used widely today is voice recognition and activation. Voice-activated boxes are one example of a technology that is changing our society. I am sure you have seen on TV or even have used Alexa by Amazon and Siri by Apple. These technologies are preparing each person who uses these devices to speak to a computer that is linked to the Internet and can understand every verbal command. Our society is being conditioned to speak to a box that can execute a command to change the stations on your radio or television without moving or answer any question you desire. Many (90 percent) of the written questions communicated to Google today will be verbalized in the future to devices provided by Google, Amazon, Apple, Facebook, and Microsoft.

There is one technology that you may not be familiar with but is key to the future of most devices in the future and widely used today. A bot is a piece of software that is designed and created to automate the kind of task we would usually do on our own, such as make reservations, add events to our calendar, or display information. It also can perform an automated task, such as setting your alarm clock, telling you the weather, searching for information on the web, managing your bank balance so you can see where your money is, providing upcoming bills, and text messaging. Even though, bot technology was created several decades ago, not until the Internet was deployed was its potential really appreciated. The real value of bots is their applications; they can go out on the Internet and do things on your behalf. It reduces the amount of time that humans would need to find information. It's like your own personal assistant that retrieves information so you can decide or not. The sister technology to the bot is the chatbot. This technology takes the information that the bot has found and mixes AI technology and now you have thousands of devices that provide you with oral or written information to your questions. Every voice-activated device is considered a chatbot. When you call the bank and ask for your bank balance, paying a bill, call a friend or relative, get directions to a specific place or location, you are in many cases communicating with a chatbot.

Your voice-activated device paying a bill with an animated picture of a person to answer your question are all chatbots. Many people call chatbots virtual assistants and think they will eventually replace call center people with these new technologies.

Now, what does this mean for me, my family, and my community? This technology in the future will replace many of the services where information needs to be provided by a human. Once the chatbot has learned most of the questions and answers provided by a human customer, they will be able to communicate the answers. With AI, this technology will better understand the type of information you want and need. We are still a few years away from full deployment of chatbots but when it's perfected, it will eliminate a number of jobs in various industries. A large number of unskilled workers to many highly skilled executives can be replaced by a chatbot.

These technological changes will be great for seniors and those who are facing physical challenges since these voice-activated devices can now turn lights off and on, give the weather report, dial a friend on your cell phone, or call for help. Unlike a pet, these devices will be able to talk to you and provide advice on your health, weight, and any other thing you want to know. A smart scale is on the market today or will be soon. It will tell you if are too fat or too thin and what you need to do. It will be linked to your refrigerator to track your eating habits and will include meal ideas based on what you have in your house. Let me give you an example of how this could work: The scale (not human) Shelia or Bob states, "Your current weight results shows you have gained 10 lbs. in the last two days." The human says, "What do you suggest I do?" Scale states, "You do not seem to be exercising enough from your blood pressure readings and the number of steps you are walking. Your refrigerator has too many high carbohydrate foods. You need to eat more vegetables and exercise more." With the ongoing development of new technologies, we will still need people who are capable of programming and managing the operation of these new mechanized devices.

Individuals will always have jobs, but they will need to be skilled in engineering, software development, application development, and cybersecurity. This latter area is the fastest growth area today with security problems at all levels.

Even 25 years ago, when Network Solutions started operating the ramp to the Internet, we had individuals who were trying to break into our systems that housed internet addresses. These hackers are becoming more sophisticated every day. This is not just a national problem, but a global one. This is the type of global warfare that can impact everything from the military defense, healthcare, credit, banking, and any other services where the information resides. It could be a real-life disruption of our national infrastructure systems of energy delivery, water, and other key systems. The more technology we use, the more vulnerable we become to security breaches of all types. You need to protect your information and question everyone that you provide your personal information to.

What Should Our Communities Do Today to Protect Ourselves?

If we wait until a breach happens, it's too late. Stay in front of these changes and develop a strategy in order to lessen the economic impact on you and your community. It is almost impossible to stop the changes from coming, embrace the changes early. These are a few thoughts on how we can stem the tide of an enormous loss of jobs and wealth if we take some preemptive action today and in the future.

1. Every home should promote entrepreneurial skills at all levels. In each home, someone should own a business. The family should support this person with resources and experience. I should not have to say this, but they should be legal businesses that do not damage the health and lives of others.

2. Promote greater emphasis in the science and technology fields with total awareness of the new advancements on the Internet, AI, and IoT. We need to read more about these technologies, and better understand their impact.

3. Teach basic economics and financial literacy to children in elementary school. Basic supply and demand theory are something that everyone should understand and how it applies to your everyday life.

4. Stay away from get-rich-quick scams. If something seems too good to be true, it most likely is. Study each opportunity and get good advice before taking the leap. This is the major reason why our communities took the financial brunt of the recession of 2008. We did not manage our most precious assets: our homes and properties during that period. The

African American community lost more than 30 percent of its economic wealth during the last recession.

5. Greater focus on healthcare and medicine should be a discipline. We need more professionals in these areas.

6. Protect and support our institutions: educational, churches, healthcare, and associations.

It is extremely important that we do not ignore these signs of change over the next decade and make the modifications in our educational and training programs. Most experts believe that the U.S. as a global player is falling behind other industrialized countries. With their youth trained for the next shift in the Internet revolution, they will be able to make job changes easily without disruption of their economies. **We are in a global race for the survival of our communities and the health of future generations.**

How will these changes impact the job market? I can't emphasize this point enough. Our community is very vulnerable to the next Internet revolution. Our jobs today are predominantly represented in the service sectors: transportation, banking, and back-office operation. Many of these functions and jobs will be eliminated under the next technology shift.

Part III: What Is Needed to Ensure Our Growth and Success?

Chapter 15:
Persistence and Determination

Story of Dr. Lonnie Johnson

There are some key attributes that are necessary to take advantage of today's fast-growing and evolving environment.

Persistence and determination are key factors in achieving anything meaningful and of consequence. This entails sometimes going against the negative opinions of others in what you are attempting to achieve. That is why I am starting this section of the book with the story of Dr. Lonnie Johnson; someone I have been associated with and have respected for a number of years. Many of you may not know Lonnie's name, but his inventions, the Super Soaker water gun, and the Nerf gun are considered two of the most successful recreational toys of the last three decades. These products invented by this African American generated more than $3 billion in revenue and substantial financial wealth for Dr. Johnson. However, it was not easy, as discussed in the brief summary of his trials and tribulations growing up and surviving in the South during segregation.

In my 30 years of experience assisting entrepreneurs throughout the country from startup to developed companies, one of the most important factors leading to success is an individual or team who has persistence and determination. They believe in their product or service and remain focused through great obstacles and even personal hardship. These are the factors that distinguish these individuals and keep them on track for success and for making more money than they could ever imagine. What sets them apart is their belief in their idea even when it has been rejected by dozens of people.

I met Lonnie in 2000 through a referral from Clarence Lott, one of Lonnie's friends from Mobile, Alabama, where he grew up. His friend asked me to visit Lonnie's facility in Atlanta, with the idea of providing him with some business advice. Most of the consulting opportunities I received were the result of someone knowing my experience and background and referring me to family members or friends. In 2000, it was a difficult year for me, since my mother, who I was very close to and my biggest supporter, died. This opportunity with Lonnie assisted me through this tough period in my life. This new opportunity commanded a high level of attention. I knew of Lonnie's toy success, but I did not know of his energy background and the work he had done at NASA, inventing energy systems for major satellite missions like Galileo as well as work on the Stealth bomber. He had a small facility in Smyrna, Georgia where he operated both his toy company and energy research lab. We had an initial meeting and he discussed his plans to revolutionize the world with his new energy technology. The first thing that came to mind was why would a person who is a successful toy inventor want to move away from the source of his wealth to a new and more risky area (it turned out that Lonnie is an energy expert first and a toy inventor second, so energy developments are not a new field for him).

I was just getting to know Lonnie and his background of grow-ing up in segregated Alabama in the '60s. One of the things I learned about Lonnie over the years, if you tell him no, that's like saying yes to him. As our relationship flourished, I flew to Atlanta each week to meet with Lonnie and his team. He wanted me to assist him in raising money for his energy technology, solid-state lithium and lithium-air batteries where he held several patents. He was seeking investors to put money in his energy projects to supplement his ongoing personal investments from toy royalties in order to continue his research and development activities. I worked with Tony Pace, his COO to develop an investment presentation for the energy company. I asked Lonnie how much ownership in the energy company was he was prepared to give up to investors and he said, "very little". I explained that because his energy project was high risk (since his business was a startup and still doing research), investors would want a major ownership interest in his company. We met with several venture capital companies on the West Coast at the start of the boom in Silicon Valley. Our list of funding organizations read like a who's who of tech venture capital firms. Many of the VCs were not interested in Lonnie's energy technology, but in the man, who invented the Super Soaker water gun. Every greeting would start off by the person telling us how many Super Soaker water guns his family owned. Then we would do the presentation on his new battery technology. In the early 2000s, no one was really into batteries like they are today. Lonnie was in front of the battery development explosion and he understood the future need for improved and smaller energy solutions. When everyone else was excited about lithium-ion batteries as an emerging technology, Lonnie was implementing a leap from the strategy of developing the next generation of technology beyond lithium-ion to all-solid-state ceramic batteries with much higher energy density. We received very little support for Lonnie's energy technology in Silicon Valley in the early days. One of the venture groups knowing my background in running several companies and working in the VC community suggested to Lonnie he would be better served by making me the CEO of the energy company, Excellatron Solid State. It would get better traction with investors knowing that I was a person from their community and with extensive business operating experience. So, Lonnie agreed, and I was appointed as the CEO. Also, with my taking over the management of the energy company, Lonnie could still be involved in the development of some future toy ideas, like the Nerf gun that is widely sold today. During this period, Lonnie relocated from his facility in Smyrna to downtown Atlanta. I was still very confused about why Lonnie was spending so much time in the energy area when he was making millions of dollars developing toys. I suggested to Lonnie that we establish a board of directors to provide some outside experience related to his strategy. One of the people that I introduced Lonnie to was Robert Holland, a very successful businessman, and the former CEO and president of Ben and Jerry's Ice Cream. He also suggested to Lonnie that he establish an advisory board and continue focusing on his toy empire.

I continued to manage the energy operation with Tony Pace and developed several investment proposals to attract capital. Our most successful source of funding was in the federal government through their Small Business Innovative Research (SBIR) grant program. This is a program mentioned earlier in my book that offers grants to technology companies that need funding for early research and development. Today in phase one, a company could receive a $150,000 grant. The other type of funding Excellatron received came from NIST (National Institute of Science and Technology) under their Advanced Technology Program (ATP) they provided a research grant of a couple of million dollars for technology for faster coating of material used in making batteries. I remember when they approved the grant to Excellatron Solid State, the director at NIST said they were providing this grant because they really did not know what we were doing, but they believed it had future value.

I continued managing Excellatron until I realized that some of the energy technology Lonnie was working on might not come to fruition anytime soon. I had been working with him for four years, living in Atlanta, and commuting to Maryland on the weekend. It was very hard on my family and I missed most of my daughter Adrian's high school years. So, I resigned from my position and continued as a consultant to the company. Today, Lonnie holds more than 100 patents mainly in advanced energy technology. His current generation of energy-related inventions began in 1998 when he licensed thin-film battery technology from Oak Ridge National Laboratory. Over the years, under his "leapfrog strategy," he has taken solid-state battery technology to an entirely new level. Whereas thin-film batteries could only be produced by a slow, expensive sputter deposition process, Lonnie's new approach uses more conventional ceramic processing techniques such as green tape casting and oven firing to make solid-state ceramic batteries. His venture into electrochemistry led him to his invention of the Johnson Thermo-Electrochemical Converter (JTEC), a highly efficient, totally new type of engine that converts heat directly into electricity. NASA is presently studying the JTEC as a more efficient replacement for conventional thermoelectric (Seebeck-type) converters presently used on nuclear-powered spacecraft. Many future energy technologies being developed today and, in the future, maybe an infringement on some of Lonnie's current patents, creating another payday beyond toys.

Lonnie's products have generated more than $3 billion in retail sales (for the licensee) in the U.S. and abroad. He is a prime example of how determination and persistence lead to success.

Lonnie turned a meaningless project of developing a water pump into one of the most successful recreational toys ever invented.

However, this was not an easy road for Lonnie, it took 10 years of persistence and determination before Lonnie gained respect for his product from the toy industry. Lonnie stated, "I had hundreds of meetings with various business groups and investors and no one believed in my product and its ability to change the face of the toy industry." It is currently ranked 20th, as one of the greatest toys ever invented.

Even as a teenager growing up in segregated Mobile, Alabama, Lonnie, a middle child with four brothers and one sister, was told he should not think about going to college by his high school advisors. Lonnie demonstrated his determination as a teenager when he designed and built a robot and became the first African American to win first place at a Southeast U.S. regional science fair held at the University of Alabama in 1964. He received the full support of his family as a young man even after accidentally igniting rocket fuel in his mother's kitchen, nearly burning the house down and getting into trouble at school with one of his experiments that almost had similar results. Lonnie gives credit to his family for their belief in his desire to become a scientist at a young age.

He continued his success and disproved all his doubters who said he should not attend college, by earning degrees in mechanical and nuclear engineering, at the undergraduate and graduate level from Tuskegee University. He was later awarded an honorary doctorate degree from Tuskegee.

He has not stopped in his quest for developing technology products. Lonnie has several products that will be major moneymakers in the future that include some of Lonnie's inventions in the energy area that will transform the global energy technology space. You can't tell Lonnie it is not possible. The idea of something being impossible motivates him to figure out how he can make it possible.

Additionally, Lonnie is giving back to his community by introducing young people to engineering and technology by building and operating robots. The robotics project has won several awards. However, I believe Lonnie's greatest contribution is by being a role model for young people, giving them hope as an example of how to become a future scientist and technologist while generating wealth for his family and community.

We have many other Lonnie Johnsons in our community who are working on projects that could change the lives of their families and communities.

Chapter 16: Reflection on African American Communities' Barriers and Suggested Solutions

Working Together

The secret for a thriving community is its ability to trust and have faith in the ability of others. Working together saves time and energy while creating wealth.

To win the race of increased economic growth and job creation in this fast-moving Internet society, we must achieve an understanding of those things that can advance our community. The individuals that have been successful in the last 25 years on the Internet creating enormous wealth were assisted by their family and friends. These trailblazers had a vision in the early '90s and 2000s of what the future of the Internet and how it could substantially grow into a business opportunity. Why did so few people see the potential of the Internet? Why are some people today able to follow their dreams and create enormous wealth for their family and community through new Internet-related businesses? This trend will continue. Most of these people would not have achieved their initial success without the assistance of others, especially their family and friends.

Jeff Bezos' family and friends in the early '90s did not understand the potential of the Internet, but they trusted in the dreamer's knowledge and experience; he gained the financial support of family members and friends to invest in his dream. Twenty-five years ago, Jeff Bezos was able to achieve his dream of what the Internet could be in the future and left his successful position in a New York investment company and drove to the state of Washington. He comments in several articles about his successful building of Amazon and how his family and friends financially supported his dreams. This support and confidence allowed him to become one of the richest people on the planet, worth $120 billion, as stated in a recent issue of *Forbes* magazine.

Many families and communities in the U.S., even if they are recent immigrants, pool their resources and support members of their community trying to provide economic wealth for the next generation. They clearly understand that it is important to support each other financially, supporting the entrepreneurial drive of others in their community: pooling resources to buy property, owning a retail establishment, contracting or supporting a technology dreamer. Their focus is on how they can achieve financial well-being as a group. Many in these communities may not even speak or understand English and have limited formal education, but they trust those inspired individuals with the experience and background to make meaningful economic contributions for their community.

This attitude is lacking in our community. We are quick to show our wealth and success as individuals, but some are unwilling to provide funding or resources to build an economic base for our community. Yes, we have difficulty gaining access to capital, but as a group, we have the resources to fund our own projects; we generate more than $1.4 trillion in spending power—more than the wealth of several foreign countries. Many new immigrant groups lack basic bank accounts but can buy property and businesses because they pool their resources.

And when they do not have the cash to contribute, they provide other resources or support to assist with the common project.

Network Solutions' name is still very prominent today, but the name is associated with a company that has been sold several times for billions of dollars. It is very difficult for me, and I am sure for the founders and former employees to forget the glory days of the company. The sale of the company in 1995 seems like it happened yesterday! Almost everyone who has heard the story of the change in ownership of Network Solutions asks the question: "How did this happen and how do we make sure it does not happen again?" I hope that my discussion of the NSI history gives people more insight into why and how it occurred. But the more important question is: did it have to occur prior to it being sold only five years later for $21 billion?

The history of African American companies not being able to reap the real benefits of a company started by someone from our community is not new. However, I have to say that our community has learned a lot in recent years from Dr. Dre's sale of the Beats head-phone to Apple for several billion and Bob Johnson's sale of BET to Viacom for more than a billion, and a few others. We need many more African Americans to develop enormously successful companies in the technology area, as well as other areas. This can happen if we work together and believe in each other. We cannot always depend on others to create jobs and economic wealth to support our future generations.

However, even with these billion-dollar success stories, there are hundreds more on the horizon like Network Solutions, technology and non-technology companies we could own. But the African American community must solve these problems today if we expect to provide a vibrant economy for our children in the age of the Internet of Things.

We need to understand that wealth today is generated by embracing future technologies and opportunities. I have spoken about this problem for several years. If we are to advance our community and build wealth for future generations, we need to learn more about where the future technology is going and its opportunities. We can learn this information by reading research reports, financial journals, technology books, and the business section of newspapers and magazines. In this book, I have included a section on why it is important to read and why the most successful people built their wealth by reading articles and reports on the direction of consumer spending and future technology. What do the experts say about the future and what are "hot" areas for development and growth?

Over the past three decades, I have been involved in assisting African American companies to raise capital. This has been a long-term problem for our community, especially when it comes to assisting other African American businesses at the various stages of growth: from startup and beyond. Yes, it is true that in the '60s and '70s we did not have the enormous economic wealth that we have today with numerous African American entertainers, athletes, and business owners joining the billion-dollar club. With this current spending power, we can easily support businesses in our community. Even with this wealth, we are still not supporting each other like most other ethnic groups. In several discussions with my associates and friends who have also viewed this issue of lack of financial support within our community, over the years, we have come to the following conclusions:

One of the major things that impede some African Americans from investing in other African Americans is the lack of trust. I thought that perhaps this was a carryover from our 400 years of slavery in this country resulting in our not really trusting each other because the "master" may be watching. We can't continue to use this as an excuse for our problems, we have invented hundreds of products and now own entertainment and sports companies. We have accomplished a great deal since slavery. But we still need to change our lack of trust and invest in each other. This is also demonstrated in our lack of support for our academic institutions that have educated more doctors, lawyers, and scientists in our community than any other institution in America. If we are to progress in this technological age, HBCUs must survive and thrive.

One person that has assisted me over the years is the chairman of Microsoft, a highly recognized businessperson in Silicon Valley, John Thompson, a graduate of Florida A&M University (FAMU), an HBCU. When I reached out to him for assistance with the modification of an agreement related to a $36 million project with Microsoft, he gladly provided advice and support. He provided support through his position to solve this problem for an African American company. Another example of John's assistance is when he introduced me to one of his white friends who was the newly appointed CEO of Motorola. He communicated with this CEO and provided my name with his support. I had a minority client that wanted to purchase one of Motorola's major energy divisions. John and I had met a couple of times, once at a reception at his house in Northern California during Jesse Jackson's Rainbow Push conference. My experience with John assisting me with my efforts is not common in our community. His success did not cloud his support of another African American businessperson. We need more people in our community who have made it up the corporate ladder providing a hand to other members of their community. John Thompson is an exceptional individual who was willing to use his influence to support me. Today, John and his wife have a foundation at FAMU assisting students with scholarships to achieve their dream.

Faith in Our Ability

Another major issue impeding the growth in our community is the lack of faith in each other. Network Solutions' story highlights this point. When the founders went to other African American businesspeople for funding or partnering opportunities, many in the community did not believe that an African American company could develop the next technology solution for the world. Over the years, I have met several bright and innovative African American entrepreneurs who have developed new software and hardware solutions years before the marketplace.

Another example is a group of engineers who came to me in the late '80s with a technology to provide broadband technology over copper wire, which was mainly used at the time for voice transmission. Major telecom (exchange carriers) companies did not believe it was possible to use copper for broadband. This group was comprised of highly accomplished and experienced African American engineers who had worked at a major telecommunications research institution for years. They were unable to receive support from the African American community and, understandably, not from the white-controlled telecommunications community either. So, they gave up their efforts after exhausting their personal resources. Two years later, the telecom industry started to introduce a new technology that provided broadband (video) technology over copper wire, the same technology for which the African American engineers could not gain acceptance or traction. This technology became the standard for the advent of video in the home and everywhere else. Another billion-dollar industry that African Americans were not able to own due to the lack of capital and faith from their own community. In other communities, investments are typically based on believing in the technical credibility of the people bringing the technology solution to their family and friends as potential investors. Most family members and friends may not know or understand the technology, but they have faith in their own people and their technical expertise. We have individuals in our community that have been trained at the top schools and held senior-level positions in the technology space who can develop the next generation of revolutionary products. But they cannot do it without some financial assistance in the early stages, from friends and family. This is the primary way most startup companies fund the early phase of their existence before they can attract outside capital. Without this funding, it is very difficult to start a business. This is even more obvious today with students being strapped with large amounts of college loans. They are already starting in a financial hole.

There are many more stories like the one above. We have a rich culture of inventors with products that have changed industries in America, but we have been unable to reap long-term economic benefits. If you do a Google search of African American inventors, you will see numerous inventions that have changed all aspects of the American and global community. God has placed in each of us the ability to achieve amazing things.

One recent experience that showed me that we have not changed our view of financially supporting each other was three years ago, when my son, Chris designed and built the first consolidated quote platform for the U.S. corporate bond market. His company, BondCliQ, now enables financial institutions to view valuable institutional pricing information for the $9 trillion U.S. corporate bond market. Given this growth in the overall size of global bond markets since 2008, this architecture is critical in supporting a well-functioning trading environment and has massive strategic and commercial potential.

Objectively, Chris is recognized as a leading expert in bond market structure and is regularly invited to speak on Bloomberg TV and at industry conferences. His 20-plus-year resume includes positions at Citigroup, MarketAxess, Barclays Capital, and Goldman Sachs. While at Goldman Sachs, Chris built the first dealer application to customize a corporate bond trading platform, GSessions, and he is listed as the lead inventor on all related patents.

For the last three years, he has been trying to raise funding for BondCliQ to further the development of the platform. Most recently, he was seeking a minimum investment of $50,000 on a total raise of $2 million. I assisted Chris by calling on my network of African Americans who have the financial capability to invest in his project. Most of the individuals I contacted also have networks of individuals who could be potential investors. Very few responded to my calls or repeated e-mails. Many in my network were former bankers, corporate executives, and board members at Fortune 500 companies. I am sure that this is not an isolated experience in our community. To be clear, the issue is not that these individuals did not invest, the issue is that they didn't even engage.

In the more than 30 years of providing business advice to our community, I have only been approached twice by the parents of African American startup company owners seeking my help for their children. This indicates to me that even today we have a problem of supporting our children. The next generation's growth is dependent on the support we provide to our young people today. Why are we not providing this support?

Many in our community exhibit an attitude that if they have made it, forget the rest of you. This is a very selfish attitude and does not help our community. We need to make a difference and reach out to help others. Some of us need to stop isolating ourselves from the community once we achieved success and reach back to help others in our community.

People in the African American community who have demonstrated experience with wealth-building need to share their experience not only at conferences but individually. Our young people are in desperate need of mentoring. We need to take the time to say: "This young person is trying to make a difference and has the credibility and the experience, let me take some time out of my schedule to give him or her a hand". Yes, many of these wealthy African Americans make large charitable contributions, but the ideal contribution that can improve our community is providing advice and or funding sources for the next generation of business leaders. I had assumed that this type of attitude had changed in our community, since the original Network Solutions era.

The book on the late Reginald Lewis, *Why Do White People Have All the Fun?* points out the difficulty that he had with raising capital. It was not until someone from the majority community came forward and recognized his ability that he was able to achieve his ultimate dream.

My son experiences reminded me so much of the difficulty Emmit and I had in raising funding from the African American community more than 25 years ago. This has not stopped Chris. To date, BondCliQ has raised close to $4 million in total capital with much of the capital coming from individual investors. Most of these investors did not know Chris prior to his pitch but eventually recognized and acknowledged his experience and background as an innovator. Very few of these investors are African Americans despite a concerted effort to engage them. If this is the experience of my son who worked at Goldman Sachs and other noted financial institutions, then it may be that not much has changed from my Network Solutions days.

Look for BondCliQ to be featured in a few years in the *Wall Street Journal* and other financial publications, providing the platform that can revolutionize the $8.5 trillion U.S. corporate bond industry and the $135 trillion international corporate bond market. This is a technology and solution developed by someone from our community.

Yes, we have numerous African Americans in our community across the country who can change our society, develop solutions, and create technological applications for the future.

We can no longer afford to allow our inventions and technology creations to not economically benefit our communities. In the technology revolution of the 21st century, our community will face substantial losses in job opportunities due to mechanization and digitized technology that we have not yet seen, but we hear about more every day. It is important that the African American community quickly puts programs in place that will provide a solution to these major problems as soon as possible, we need to solve this very serious problem of trust and faith in our community. When I created my presentations 25 years ago trying to convince our community to become owners on the information superhighway, we did not know the potential of the Internet. Today we are one of the largest users of Internet services making billion for others, especially social media (Facebook, Twitter, Instagram, etc.). Minority communities are also major users of smartphones in the country. However, we are not benefiting from these technologies economically except as "riders" on this fast-moving technology highway.

With more successful ventures in our community, generating new capital to solve these problems, we will be able to create a large resource of funding for other investments. This will allow us to reap the long-term benefits of a company like the original Network Solutions and not just give it away.

I don't want to belabor the problems of our community and its inability to support African American technology development; I want to provide a couple of solutions that we should consider. One point that I would like to emphasize before providing additional thoughts is that we have an amazing population of young people with great ideas for generating new products and services in this new technology revolution. Additionally, through STEM programs, we are training many more to follow in their footsteps, but we must provide them with financial resources to develop their dreams and ideas.

What we need to do immediately is to identify those individuals in our communities with the resources and desire to assist in providing capital for funding technology and inventions created by African Americans. The venture capital and private equity communities, which are predominantly controlled by others, have not provided employment opportunities for African American young people in the industry. This is an industry in which we still represent less than 3 percent of the managers and traditionally receive less than 2 percent of the capital from these sources; this is not expected to change much over the next few years. However, we do have several experienced investment bankers and investment advisors within the African American community who we need to support in capital formation initiatives. In the last couple of years, a number of wealthy African American businesspeople, entertainers, and athletes have demonstrated their faith in our community by making investments in our education and small business startups. We have organizations like the National Association of Investment Companies (NAIC), an organization of minority-owned venture capital firms with $150 billion in resources playing a major role in providing financial assistance to minority-owned companies as well as business management advice. This organization has highly experienced and accomplished managers who need greater financial resources from major financial institutions and wealthy individuals in our community. Another organization that has also played a role in the matching of capital sources with high potential minority-owned businesses is the Culture Shift Lab organization located in New York They provide a bicoastal gathering of entities with capital resources for rising technology dreamers. Many of these entrepreneurs would have been unable to access these funding sources and wealthy individuals without Culture Shift Lab's annual networking event in New York and Silicon Valley. We need more of these types of organizations to bridge the capital resource gap in our community.

While completing the writing of this book, a historic event occurred on Sunday, May 19, 2019. Robert Smith, one of the most successful African American businesspersons and members of NAIC, announced at Morehouse College's graduation that he would pay off $34 million in student debt of the graduating class of 2019. He had already donated $1.7 million earlier in the academic year. Robert Smith is the owner of Vista Capital, one of the most successful, venture capital companies in the world. Robert is worth more than $4 billion and has shared his wealth with numerous African American organizations, minority STEM programs, and was a major donor to the National African American Museum, among others. Robert is also an alumnus of Columbia Graduate School of Business. He and his parents hold doctorate degrees from the University of Denver.

Obviously, we do not expect other African Americans to follow Robert Smith's example unless they have this type of wealth, but we should expect others to offer greater financial assistance to the HBCU community. Let's not ignore one of the most valuable assets in our community.

Another person committed to strengthening the educational system in Akron, Ohio is LeBron James, the famed professional basketball player for the LA Lakers. He funded a public school in the community in which he grew up. He is providing his financial resources to assist families by providing a quality education system and promoting their advancement to college. This should serve as a good example for other NBA players and athletes to provide financial resources to improve the health of the educational systems in their communities.

There have been a few noted athletes and entertainers who have publicly stated their commitment to providing capital to startups and growing enterprises. My hope is that many of their dollars will be directed to dreamers in our community. We are the ones who desperately need financial assistance and mentoring. Some of the individuals who are investing their dollars today are Serena Williams, Venus Williams, Kevin Durant, Kobe Bryant, Steff Curry, Will Smith, Shaquille O'Neil, Carmelo Anthony, Jay-Z, and Ludacris.

I hope that these individuals, after reading my book about the history of the Internet, one of the most powerful technology tools in the world, understand that African Americans were instrumental in bringing it to the global community. The Internet has changed the world forever with more than 4.2 billion users today and an economy estimated at $8 trillion, as stated by McKinsey & Company.

Understanding Risk

One major point I would like to make is related to investment risk. Most experienced investors do not expect all their investments to be successful and to generate a return. In fact, it is generally only 20 percent to 30 percent of the investment portfolio that supports the investment fund to make returns large enough to offset losses. No one can expect that every investment in the African American community or any other will be successful. Risk is of major concern in making an investment in a startup or any new company. That is the main reason why the backers of startup companies require a major share of company ownership when they write the check.

In most successful startups, both tech and non-tech, someone took the early risk in providing the funding. Most angel investors (less than $250,000) and VCs are generally not making one investment, but multiple ones. The reason for their strategy of placing their bet on several entities realizing that not all their investments will be successful. In fact, most of my experienced investors do not expect on average more than 30 percent of their investment to be successful. The successful ones generally make enough money to cover the losses.

We need to look at risk with the expectation that this next investment may be that major success. However, all you need are a couple (like Network Solutions that sold for $21 billion) to be adequate to cover your return. For example, SAIC stated in their books that they generated more than $3.2 billion in cash from their Network Solutions transaction that cost them less than $5 million in five years.

Many of the African American investors mentioned in this section of the book have taken some risk on various projects and lost money. However, they continue to invest their money seeking those successful opportunities, so that they recover their losses and make more money.

Jason Feifer, the editor of *Entrepreneur* magazine states, "In taking a risk to achieve a vision, we can change the rules and create lives of our own making. Risks don't always work out, but they do always remind us of the vast possibility waiting for us."

Asking the Right Questions in Evaluating an Investment or Loan

From my experience, I would suggest that you get the advice of an attorney when you are ready to make an investment or loan. You need to document the funds that you are providing and know what the expectations are for the investment or loan. The key issue in making this type of investment or loan: Can you afford to lose the money without having a major impact on your life. Also, is the person you provide the loan to someone you can trust?

I have found that family members and friends who are asked to make investments or loans for a business opportunity decline mainly because they did not know what questions to ask about the potential investment or how to assess the risk. The business does not have to be in an area or endeavor that you totally understand. In fact, there are multiple technology areas that most people do not understand that have a high potential to grow and make money in the future. In the early days of development, Amazon, Facebook, Google and other technologies that we know of today, many investors did not have the vision of its founders. However, they believed in the people that ventured into these new areas of technology and had faith in their vision. Most of the early funding came from family and friends, who today do not regret taking the risk for a family member or friend.

I have prepared a list of questions that will solicit critical answers in your decision-making process. These questions will also sharpen the skills of the person making the request when they meet other potential investors. These questions will give you a basic understanding of the business risk and that the person has done his or her homework in developing their plan and all potential problems. Remember that you are not attacking the person's idea but ensuring that they are prepared and have an in-depth understanding of what they are doing. You want to solicit firsthand information from the person making the request for financial help.

When considering making an investment in a family member or friend's business these are some questions that you should ask to help in your decision. These are not all the questions that you need to have answers to, but some basic ones.

1. How much money do you need and how do you plan to use the money?

2. How much research have you done on the potential opportunity and how long will it take for me to be repaid?

3. Do you or your partners have any experience or skills in this area?

4. Who are the competitors in your area of business and what share of the market do they have today?

5. Do you have some special advantage that will differentiate you from your competition?

6. How large is the market for your product or service?

7. When do you expect to make a profit?

8. What percentage of ownership are you offering me for my investment or loan?

9. Who else has invested or loaned you money for the business? What have you personally invested in time and money?

10. May I see your business financial projections or business plan?

Learning from Failure and Disappointment

Many people do not take on the risk of owning a business because of the fear of failure. There are many people who are successful today because of their failure in one or more businesses. The learning experience of failure(s) helps you achieve an understanding of what works and what doesn't work. It is rare that a person is successful in their first business venture. Those who are successful in their first venture may have a difficult time repeating their success since they have not learned from issues, they faced in building a business. That is the main reason why people who have made a lot of money in their first venture often lose it quickly. They think that lightning will strike twice, and, in most cases, it does not; then they wonder why they are not successful the second time around.

I have had clients who made money easily in their first venture, lost all they had and could not get back to where they were. If you are lucky enough to have success in your first venture you need to manage your money cautiously and don't go crazy with your success.

Billionaire Bill Gates stated, "It is fine to celebrate success, but it is more important to heed the lessons of failure." Many of Bill Gates' projects were failures when he and Paul Allen first started Microsoft. From their failure in coming to the market with the wrong product in their early stages of development, they realized what would work and made millions and then billions from their mistakes. Jeff Bezos had many failures and lost a great deal of money until he realized that some of his ideas didn't work. He kept learning from his mistakes and taking chances.

With the increasing number of people wanting to build the next Facebook, Instagram, Google, Amazon, PayPal, etc., many assume things will just happen. Most new entrepreneurs do not realize how difficult it is to start a successful company and keep it going day-to-day. When I started my first company, I thought due to my previous successful banking career, it would be easy to raise funding. I completed a loan application at the bank that I had just left. The commercial lending officer stated to me, "Al, you were a great banker, but you have no experience in making payroll." I was very upset that I was turned down by the bank where I was head of the International Finance Group. What was he talking about, never made a payroll? Very shortly after receiving this answer I learned that the commercial banking officer was correct. One of the hardest lessons for a new business owner is to learn that when you have employees, the responsibility of paying them is on you. Owners of new businesses are generally the last to get paid, but that's the cost of being the major shareholder.

When you have been disappointed by rejections from investors or bankers, just don't walk away pouting and upset. Listen closely to their feedback and why you were denied. This bit of information may help you when you make your next presentation for funding. Many successful businesspeople have been rejected many times. They get off the ground and keep pushing but learn something from each experience.

Many entrepreneurs believe that their idea is great and will make a lot of money. A dream without a plan is just a dream. I have reviewed a lot of projects that were ideas that could make money, but the critical issue that dreamers failed to consider was how were they going to execute their plan to achieve the potential of this great idea? What resources and assistance did they need to make this thing happen and keep it going? During many discussions with inventors or developers, I would have to bring them down to earth on the potential of their project. Yes, they have a patent or pending patent, but that does not guarantee the success of their product or idea. I have seen proposed projects where the scale of the project was so large that it would cost hundreds of millions to even make a profit. This is not to say that the project must make a profit initially. In fact, Jeff Bezos did not make a profit for more than seven years, but he still was able to convince investors to fund him. Some newbies have an idea for something that already exists in the market and think they can improve it, not realizing that their solutions may already be in the works with the company or individual who had the initial idea. What I suggest in these cases is that they contact the primary developer of the product and seek a partnership or licensing agreement. Several of my clients stated to me during our first meeting, that they did not want me to know what they were doing, fearing that I would steal their great idea. I asked them, "Well, how are you going to raise capital if people are not allowed to know and see what you are doing?" I generally get silent treatment and very few words of discussion since they have not thought through that problem.

Was I scared when I started my own business? Of course. I even questioned if I had left banking too early since many of my friends from graduate school were still getting a check every two weeks.

I made several mistakes because of my lack of experience with a mentor. Even with my business consulting experience in providing advice to established businesses in Harlem and Denver, I had not started a business of my own. I learned a lot during those years that are the basis of some of the advice I give startup companies today. The first lesson I learned is to keep your expenses low and do not expand until you have regular money coming in the door. When I started my first business, I rented a plush top-floor office at 17 Battery Place right off Wall Street with floor-to-ceiling windows. I had a grand opening and invited all my friends who marveled at the size of my office and the view. I had relocated to this office from my apartment because I believed that clients would walk in the door with business and be impressed. Well, no client gave me a contract when I was in that office and eight months later, I was back working out of my apartment on the Upper West Side of Manhattan. In fact, when my office was in my apartment, I earned more money and signed more contracts during that time. You do not need a plush office to generate business unless you are a doctor or lawyer and even then, you should start small. I think the incubator and space-sharing office boom for tech companies today is a great way to save money and stay in business.

There is always a level of risk when starting your own business and failure can only make you smarter along the road to success. You need to be persistent and not let failure stop your efforts. Now, if you continue to make dumb mistakes and not gain insight from your experiences maybe you need to work for someone, instead of your own business.

Since leaving banking, I have started more than half a dozen companies; with each experience, I learned some valuable lessons.

One of the companies I started was marketing a unique promotional product. It was a product that allowed people to enjoy concerts or any viewing event with their own portable binoculars that could fit in their pocket. The product was made of plastic and you could print any image on the product. It was considered a "walking billboard" and some of my friends today still have some in their drawers at home or in their office. I spent a considerable amount of time marketing and selling the product to record companies to promote new records, beer companies, and just as plain old binoculars. During the time I was marketing the product, I received a preliminary contract from Pepsi to buy more than one million units for a Michael Jackson global tour; they planned to give them away. I had been looking for a contract like this for years and I had one that would launch my product to make millions. Pepsi wanted me to travel to Los Angeles for the filming of the commercial for the Michael Jackson tour. I had a backstage pass and I stood next to Janet Jackson watching the filming of the commercial at the Shrine Theater. Music was blasting and dancers, as well as Michael, were on stage, and the smoke machine was blowing smoke to open the commercial. People in the audience were screaming to the music of "Beat It." Michael walked around the stage when someone screamed that his hair was on fire and a stagehand rushed in with buckets of water to put out the fire. The Pepsi executives were going crazy and Janet ran to Michael to see how he was doing. The next thing I knew the commercial was canceled and so was my contract. Pepsi was more concerned with the legal action by Michael's attorney than my contract. I called my wife who said that the TV news was showing the picture of Michael on fire. I never got the chance to get my contract signed and chalked up the incident as "Murphy's Law" changing my successful opportunity to a disastrous situation I had very little control of. So, you always must expect the unexpected when you are in business.

Getting Good Advice

There is no substitute for experience, no matter how smart you are. Most people who are new to the business and are trying to launch a product or service will seek guidance from the wrong people. I was approached by a company in California that had great wearable products for women. They wanted to know if I could help them raise capital. Following my initial review of their product and strategy, I realized they had several holes in their plan more than a lack of capital. One of their issues was their inability to find a reliable manufacturer in China. One of the partners had prior experience, but limited manufacturing knowledge for this type of product. I was really concerned and introduced them to a friend of mine in Taiwan. He had been successful for more than 20 years with extensive experience working with Asian manufacturers; his name is Jason Ray. The bigger problem was their marketing strategy. I suggested that since their product was being directed to women, that we identify an entertainer or sports figure to help brand the product. One of the other concerns was that their product had a low barrier to entry, meaning that anyone with money could enter the market with a similar product. Timing is everything, but they thought that many of their problems could be solved with more capital. They missed the point and the market because they failed to listen to experience. This is a lesson that many new businesspeople learn late, rather than sooner.

One of the major issues that prevent businesspeople from becoming successful is not retaining an experienced person to assist them. Many newly formed businesses and their founders are unwilling to pay an experienced professional to assist in helping to execute their plan. This generally is a costly mistake in the end. They will be told by friends; you do not need to pay for advice in executing your plan or even raising capital. When friends or family members say that to you, then ask yourself what they know beyond being addicted to one of the popular investment TV shows. Remember you are in a business where you have hired attorneys and accountants to help you in their areas of expertise. You should think of the business consultant and advisor in the same way. They have many years of experience and can assist you in avoiding the pitfalls that other businesses have succumbed to.

One of my former clients asked me for assistance in helping his company penetrate the federal marketplace. The company had no prior experience in the federal sector, but it had provided services for major disasters and had managed a large sum of money during Hurricane Katrina. They wanted my expertise and experience but initially did not want to pay me market rates for my services. Finally, they decided to allow me to develop and execute a plan to get them a federal contract. Even though they were a minority-owned company, they were too large to participate in small business programs. We agreed on a budget and I opened an office for them in Washington, DC where I directed the marketing program for opportunities. Since I had extensive contacts and understanding of the federal regulations, I investigated opportunities related to their capability in several federal business areas. While I was executing my plan and strategy, they were still not paying me the full compensation that the position and responsibility deserved. I showed them a salary report from other people at the senior vice president level in the federal sector. During my negotiations with the company, I won them a contract with a small federal agency, that generated more than $28 million in revenue over five years. The owner of the company did not believe it would happen in less than 12 months and mentioned to me his surprise. His managers that lacked any federal experience filled his head with doubts about my ability to perform. Even with this win, the owner still had a problem paying market rates for my services and I resigned from the company. The company was unable to duplicate my success in other federal opportunities. But they tried to do it with inexperienced people. This company had the potential of generating a substantial number of federal contracts, but they were not willing to make the investment in an experienced person.

This problem can also exist with paying a consultant or advisor a retainer. I cannot afford to pay you until I see something. Many companies lose out on major opportunities because they are narrow-minded in not using consultants and paying for the service. If you do not know what you are doing find a tested consultant or advisor who can help you.

Communication and Knowing Your Strengths

Communicating effectively is paramount in the business world. When I started my own company following my career in banking, I was forced to write more and speak in front of more public audiences. As a consultant, your ability to provide your clients with a concise written document of your proposed strategy and plan is necessary. We have the tendency as individuals to verbalize more than necessary than in writing. When you are trying to raise funding for your project you need to write a detailed business plan and provide financial projections. We have departed from the traditional 50 to a 100-page document to explain your project and the background of your team. Most business plans today are a short business summary of the opportunity and how you plan to spend the investors' money. Investors generally can determine in the first 30 minutes of your presentation if they have confidence in the project, but more so in the team, to deliver on their promise.

I learned over the years that there are great products and services that do not meet their full economic potential because the people were not the right fit to execute a winning plan. Usually, they knew too much about the product or service but did not know how to execute the plan. One big lesson I learned early with my own companies was that, in selecting partners, even if the person has a great resume and good educational credentials, it does not make them a good partner or manager. Feeling confident that you can trust the person is more important than the number of degrees that they may have. In fact, I allow my wife, Gwen, to meet people following my interview, to see what she thinks. Women have a greater sensitivity to the personal character of people than most men.

One major rule in hearing someone speak about a problem or issue related to unfulfilled promise is this: if a person must repeat something three times, they are most likely trying to convince themselves, not you. Let me give you an example: I asked one of my managers if he or she planned to take care of a critical task. They stated, "Don't worry, it will get done." Once they said that three times in less than five minutes, I lost total confidence that it would ever happen. When someone borrows money from you and they say three times, "I will get the money back to you next week," don't expect it to happen. This is my three times and it won't happen rule.

Use of Non-Disclosure Agreements

Most new companies believe that they can protect themselves if they get an NDA signed by the people receiving the information. The Non-Disclosure Agreement (NDA) is rapidly losing acceptance in the VC community and the general public. No major financial organization wants to be subject to legal action before they know what you have to offer. So, most venture capitalists do not sign NDAs but ask for initial information that is non- proprietary. Remember, you need their financial assistance and they can walk away if you seem like a difficult person to work with. Investors see hundreds of deals and hear numerous stories of how great a product or service will be in the future. More than 85 percent of an investors' decision to write a check is based on the personality and chemistry of the team. If the investor feels unsure of the team's ability to deliver on their plan, they walk away. You may have a great product or idea, but if you can't convince the investor that you can achieve your goal and provide them the return you projected; you will be unsuccessful in gaining their financial assistance.

Chapter 17: The Attribute of Billionaires: Reading

While writing this book, I thought about how one of the most important aspects of achieving personal success and wealth in any person's life is to have a commitment to reading and gaining knowledge. Most successful people are lifetime learners.

Regardless of how many degrees you have or do not have, expanding your ability to understand what's happening today and in the future will be very important to your growth and development; it may even lead to your personal wealth.

When I was attending the University of Denver's business school, my professor of economics said that you must read if you want to know what's happening. This advice I have cherished for most of my adult life. For more than 30 years, I was a subscriber to *Business Week now Bloomberg*. Each Friday, I would wait anxiously to read what business-people around the world were saying about various issues affecting the global economy. It also provided a look at the future industries and trends that I could capitalize on in my consulting business and positions in banking, technology, energy, and healthcare. It also provided valuable information about foreign markets and trends abroad. This information gave me the tools to better understand which industries and areas of business I should track.

Philip VanDusen, founder, and owner of Verhaal Brand Design states the following: "Books help us in so many ways to succeed as an entrepreneur. They can be a source of inspiration, develop skills, and provide tips and business strategies. They can help us to become successful by providing the right knowledge, new ways of thinking, new insights, and help us develop the necessary skills. Reading can also influence the way that we do business through personal development and leadership skills. We can get a lot of new ideas and tips allowing us to have multiple perspectives to connect the dots and do things in a way that had never been done before. By reading books we also develop cognitive skills that can benefit us in a multitude of ways by helping to improve our memory, develop literacy, and verbal intelligence as well as increasing our strategic decision making, boosting our brainpower, and reducing stress. Not only can we learn from the successes of other entrepreneurs, but we can also learn a huge amount from the failures of other entrepreneurs. It is very important to make sure that we learn about both perspectives."

When I was a banker at JPMorgan and Bankers Trust, I believed that I could surpass my peers, since I knew more about the global economy and the various statistical indicators. Additionally, I read the *Wall Street Journal* that provided daily information and *Businessweek* now published weekly by Bloomberg. Other publications that have been extremely helpful, even in writing this book, are reports published by Nielsen Company, McKinsey & Company, and *Wire* magazine. They provide global data analytics and research on the various consumer market trends. They especially do a good job of providing statistics on the minority markets related to buying habits in various sectors. The other company that has great research and data is the Pew Foundation, a nonprofit company providing outstanding information on the trends related to the Internet and Nielsen Research with similar information. Their Internet reports relating to social media trends in the minority community are excellent. Some of their charts are included in my book so you can get a glance at the current and future trends in Internet usage. This type of information is very important when you are trying to develop a business plan focusing on the size of the African American and Hispanic markets. The total minority market annual purchasing power exceeds more than $3.5 trillion (2018), more than the GNP of several countries in the world. All this information is available online generally at no cost. However, you should consider donating to their organizations after you have made your first million by utilizing their research.

In the past I would retain various weekly and monthly publications for months, believing that I needed to hold on to them because of their valuable information.

One other publication online resource and that I cherish is *Black Enterprise* magazine that reports on what's happening in the African American business community. The publisher of this renowned and informative magazine/online services is the late Earl Graves Sr., another famous graduate of Erasmus Hall High School.

Black Enterprise published an annual list of the top 100 African American businesses. I collected these issues for a number of years, so I could track the leading African American companies and their specific industries. The original Network Solutions was listed several times on the *Black Enterprise* top 100 lists in the '80s s. One of the astounding facts is that during one-year *Black Enterprise* top 100 listed three companies managed by CEOs who were graduates of the University of Denver. This was amazing to me because the University of Denver had a relatively small African American student population in the '60s and '70s.

Access to magazines and newspapers of all kinds is much easier today with Internet technology. However, I still like to read hard copies of many of these business publications.

I also buy online audible or e-books on business and technology-related topics.

Some of the most successful businesspeople in the world stated in a Waqar Ahmed article in the *Business Insider* that Warren Buffet, one of the most successful businessmen living today and (also a graduate of Columbia University) reads more than 500 pages every day. Buffet states, "That's how knowledge works. It builds up, like compound interest." Bill Gates reads 50 books per year, which breaks down to about one book per week. Elon Musk is an avid reader and when asked how he learned to build rockets, he said, "I read books." Mark Zuckerberg forced himself to read a book every two weeks through 2015. As you may know, Oprah Winfrey selects her favorite book every month for her book club members. Most of these individuals gained their insight into making money by reading about other high achievers. Successful people don't just read anything. They are highly selective about what they read, opting to be educated over being entertained.

One person that I have been fascinated with over the years is the late Nikola Tesla, one of the greatest scientists of the 1900s and of our time. If you are studying any aspect of STEM, you should become familiar with this Serbian American engineer and physicist. Tesla made dozens of breakthroughs in the production, transmission, and application of electric power. He invented the first alternating current (AC) motor and developed AC generation and transmission technology. Some of his greatest inventions ushered in the use of remote controls, such as the device you use today to change channels on your TV. He perfected the use of the x-ray for detecting body parts and images and the neon lamp that we use today for lighting in most public places.

He was light years ahead of most commercial technology development. Most of his technology patents were sold to Westinghouse Corporation. When he died in 1943, he was penniless and discovered in a one-room hotel in New York. His inventions provide all sorts of benefits to our everyday life. What fascinated me about Tesla is his history as an inventor and his life mirrors the life of most early African American inventors during the last century by developing technology that has changed our society forever, but not reaping the financial benefit since they sold their rights in order to survive. However, the only African American inventor to have his name associated with his invention "The Real McCoy" is Elijah McCoy, for the railroad train braking system, whereas Tesla's name is associated with the Tesla electric automobile. I am sure many of my readers may not be aware of that fact.

Chapter 18: Message to African American Millennials

One of the reasons that I wrote this book was to share my business experience with the African American community and provide some direction in this technological age of the Internet. However, there is one group that I especially wanted to reach and provide some guidance to in navigating future opportunities: African American millennials.

Many people do not realize the enormous size of the millennial population today; they surpass the total number of baby boomers (born between 1946 and 1965) of 74.5 million compared to 75.4 million millennials (born between 1981 and 1998). This chart below points out the projected growth of all generations by 2050.

As the baby boomer population in their heyday, millennials are the focus of major marketing campaigns due to their enormous market presence and economic wealth potential. Many in this group had retirement plans and spendable money for many years until the dotcom bust of the early 2000s and the housing crisis of 2008 that substantially reduced the asset wealth of the baby boomers, especially African American baby boomers.

Since I was born at the beginning of the baby boomer generation and experienced several technology shifts and economic changes during my life, I wanted to share some of my thoughts with African American millennials, and all millennials in general.

Projected population by generation

In millions

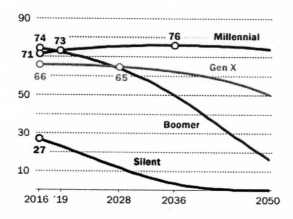

Note: Millennials refer to the population ages 20 to 35 as of 2016.
Source: Pew Research Center tabulations of U.S. Census Bureau population projections released December 2014 and 2016 population estimates.

PEW RESEARCH CENTER

One of the lessons that my parents drilled in me most was, "Save your money and reduce your debt." This suggestion is more important today than ever before. Many individuals during my generation were not frugal with their money and were more concerned about the size of their house and the car they drove, the image. The baby boomer generation earned good wages and sales/management positions in corporations in the '70s to '80s; we had the advantage of a stable economy and the buying power of the dollar went a lot further than today. I am sure you have heard other baby boomers talk about having middle management positions in the late '60s and '70s and living well off a salary of $12,000 to $20,000 a year. Today, these salaries would be considered below the poverty level. However, during that period you could buy a fancy car for less than $4,000 and rent a big apartment for $300 to $500 per month. And we still had plenty of money to travel.

Life was pretty good, and we conserved some of our money in the bank, but not enough. What has happened over the last 30 to 40 years is that wages have not gone up as fast as cost-of-living expenses, meaning the dollar does not go as far as it did. Today, basic salaries are not rising as fast as increases in the cost of housing, food, and other living expenses. Additionally, most baby boomers who were college-educated were able to pay off their student loans in less than five years as opposed to the 10 to 15 years it takes for graduates of today, if at all.

The economic environment in the future will be much more focused on wealth and survival than it has been in the past. We used to think if you made low six figures you were doing really well and could live a good life. However, if you are living in a major city like New York, Washington, DC, San Francisco, Chicago, Boston, Miami, and several other cities, with a low six-figure salary, you may be just getting by. It is so amazing to me how we talk about the number of billionaires, not millionaires today. Everyone is waiting to see who the first trillionaire will be. That's how crazy our society has become today with our focus on wealth, not health. The wealth gap between the wealthiest people and the poorest is getting larger.

Things are changing so fast related to technology, as already mentioned with digitization and robotic advancements, you must be ready to make the shift before it comes and have the resources to survive. There are already discussions in Washington, DC about limiting future unemployment benefits and social security. The birth rate is slowing among the non-Hispanic white population and African Americans. The only population that is growing in the U.S. is the Hispanic population. With the new immigration laws, we are likely to lose this base of contributors to the social security system, which would have benefited millennials 30 to 40 years from now when they are ready to retire if they can ever.

One economic principle to remember that other generations failed to understand during major economic growth periods is believing that the good economic times will be here forever; that just doesn't happen. History has shown us that economic growth is cyclical, you have high points and low points. Good examples of this economic principle were the housing bust in 2008, after several years of rising housing prices (eight years), the dotcom bust in the late '90s and fall of 2000 and the recession of the '30s. When things seem to be good for five to 10 years, look for the correction in the market. Records show that there have been eleven recessions in our country. While I am writing this book, we have seen more than ten years of tremendous growth in stock prices beyond normal levels. This period reminds me of the Dotcom bubble several years ago. Inflated stock prices and valuations with newly established public companies facing the problem of generating a profit.

Most astute economists today are already projecting a near-term cyclical correction to take place with stock prices and other assets falling. Most very wealthy individuals in the country and around the world know this fact and are sitting on large piles of cash waiting for the downturn so they can buy into depressed markets. This is how they made their wealth, by buying at the low point of the market. Just remember this principle when you are building your business or empire. Therefore, it is important to have employment options and savings.

Millennials must look at the technology and social media as the means to improve their economic status over the next decade. It should not be taken for granted, as being only a means for knowing what your ex-boyfriend or ex-girlfriend is doing, or who said what to who, or whose picture is being posted on your Snapchat or Facebook page. Think of the Internet as a medium to communicate, like the radio, TV, and movies. Over the years we have learned how to write and produce music and entertain audiences as comics, singers and interviewers. Professional athletes from our community similarly have benefited over the years from increased dollars flowing through the communications industry and providing enormous wealth. Today, African Americans own music and film production companies and benefit from everyone who buys their products at all levels. The African American community has been blessed by God for more than 200 years, as the ultimate entertainer. We are admired all over the world for how we entertain the public and even for successfully leading a super-power country for eight years.

Social media and the Internet should be looked at the same way; you can make money providing products and services through this expanding technology, as the examples I have shown earlier in my book.

It is much easier today to showcase your talent and abilities to not just millions, but to billions of people around the world. Imaginative individuals have become very successful in showcasing their talents on YouTube or through podcasting. The more creative you can become; the more opportunities will be available to you. One of the fastest-growing markets today and in the future is providing entertainment (talk shows, skits, movies) for people in the developing world: Africa, Latin America, the Caribbean, and parts of Asia. However, what is necessary to provide these platforms in other countries is the ability to speak their native language. Most of the information on the Internet is in English or Spanish. However, there are billions of people in many of these countries who do not understand either language. Billions of people that are not reached by the Internet today will soon be linked. One country that has taken advantage of this current and future growth in Nigeria. They are producing their own YouTube soap operas, thousands of shows per week for their market in Nigeria and for millions of Nigerians that live outside their country. So, millennials and others who are from other countries, start thinking about producing your own shows that can be broadcast internationally.

Also, millennials need to start developing Internet platforms that take advantage of IoT and AI. There are hundreds of startups developing labor-saving ideas and the means to operate things remotely with very little human participation. One of the things I personally plan to do is to take an AI class to better understand the technology and how it will impact me. This is an area with which every millennial should become very familiar. This is a technology that will influence most decisions in the world for most of your life. Understanding what you do with this technology will provide enormous business opportunities. Get in front of technology; don't wait. Last year I met some female African American millennials operating in Brooklyn who had designed clothing with sensors sewn into the fabric. They had created an inventory and security tracking system for retail stores and manufacturers. The sensors can track the number of units sold instantly. I am sure they are doing well today.

Getting an experienced advisor when trying to grow your business and capitalize on tracking what is new and innovative. Look at the AngelListblog.com site if you are looking for a tech position with a fast-growing startup or seeking capital. This is perhaps the best way to learn, by working with a startup and participating in the building of the business. Most successful tech companies today were formed by an entrepreneurial team that helped make a tech company a lot of money and then individuals from the team left with their experience and lots of money. Don't be afraid to take a low-level position with a startup. There are numerous secretaries and analysts who are millionaires today because they started at the bottom with a successful startup, like Google, Facebook, Amazon, or Uber. The purpose for taking these positions is to learn and understand how to grow a fast-moving technology company and generate your own capital through stock options to fund your company. Startups across the country are having a tremendous problem today finding people with basic skills in marketing, sales, computer science, and other critical areas. This is becoming even more of a problem because of the changes in U.S. immigration laws forcing engineers and technical people who are not naturalized U.S. citizens to return to their home countries.

As stated by the Nielsen research report, there are more than 11,400,000 African American millennials today, a little more 25 percent of the African American population. More than 90 percent of African Americans millennials own smartphones and 86 percent accessed the Internet exclusively on mobile devices in 2015. Marketers of consumer products and services consider African Americans millennials today a major target market because of their savvy use of technology and the ability to educate and teach others about current Internet technology.

The Nielsen data suggests 55 percent of African American millennials spend at least one hour on social media sites, which is 6 percent more than all millennials. Nearly 30 percent spend at least three hours on social media sites, which is 9 percent higher than other groups.

If you want to know more about the buying habits of the African American Community at all levels, I would suggest reading the Nielsen Research Report: https://www.nielsen.com/wp-content/uploads/sites/3/2019/09/2019-african-american-DIS-report.pdf

Millennials are major users of Twitter and without their active participation, Twitter, as well as other social media sites, would have to close. African American millennials represent a major portion of our communities the $1.4 trillion annual spending.

African American millennials are positioned to drive future economic growth of our community through this technology revolution with their extensive knowledge about the Internet and social media than any other group in our community. You grew up with Facebook, Google, Instagram, and Twitter, and other technology companies. Now, after your extensive experience with social media and search engine sites, you are proficient in most of the Internet platforms. You demonstrate these skills of using the Internet with your call to action across the country when African Americans were being murdered by local police. You have established yourself as a social power. Most people do not realize that millennials during these protests were able to gain the financial support of multiple organizations and wealthy individuals across the country. The movement generated more than $150 million in funding for their movement. Most of these donations did not come from our community.

You now need to take this same energy and brainpower to start developing opportunities on the Internet to build economic wealth for our community.

This same power and knowledge can happen on the Internet. We just need to stop for a moment and consider ourselves as much more than consumers, but as owners and producers. The Internet is the largest network in the world with billions of people looking and listening each day. We need to create sites and programs so that these billions of Internet users can impact money flow into our community. I think you can play a major role in developing programs to direct this $1.4 trillion spending power in our community to more of our own businesses or companies. There are still more than three billion people that do not have access to the Internet mainly in the emerging markets of Africa, Latin America, the Caribbean, and parts of Asia, who can become customers for your service in the future.

I am sure that African American millennials can achieve the goal of the original Network Solutions in making history by developing a product or service worth billions of dollars.

Chapter 19: Looking Forward by Building on the Legacy

Since December 31, 1992, when the original Network Solutions was designated by the National Science Foundation to provide Internet addresses to the global community, many positive things have happened to the founders and others associated with the company. The story of Network Solutions and its founders not being able to fully capitalize from the two public offerings and final sales of the company for $21 billion is not the end. This story coming to light over the years has been a learning experience for other African American companies with disruptive technology to follow their dreams to the end. Do not allow others to tell you that your vision and insight in a particular direction is not worth pursuing! Many organizations or people that reject your ideas do not have the vision or understanding of your potential. Remember that when I tried to promote the use of the Internet in the '90s, very few understood the technology and the benefit of getting on board early. However, there are a few individuals like billionaire Jeff Bezos, who had a dream and vision of the potential of the Internet being a global marketplace to sell any product or service.

The next generation of Internet technology has just started and offers a vast number of new opportunities to developers and inventors. Every setback is preparation for a greater opportunity, if not for this generation, then for future generations.

The founders and employees of Network Solutions were trail-blazers breaking new ground for technology in the 20th century. Their accomplishments can't be diminished. There is only one, first-place award for winning. Network Solutions was the first to bring Internet technology to the global community.

The other winners are the families of the founders and employees who were a part of the accomplishment and learning experience. Passing on these experiences is very important to the growth of our community. Our children and grandchildren can benefit and achieve great things in the future from hearing and seeing examples of our successes and failures.

The legacies of the four founders of Network Solutions have already resulted in the success of their children. Emmit McHenry has two sons who are highly successful businessmen today. One, Kurt McHenry, is the COO of a telecommunications company in Virginia, and the other son, Michael, is a record producer and writer on the West Coast. Gary Desler has three sons. Two of them are business owners, and the other son is a manager and works in the restaurant industry in Maryland. Ty Grigsby has no children but mentored a young man with whom he jointly created an international entrepreneurial program. They set up training facilities throughout Africa; it is called the Nehemiah Program. Ty has also been a member of his church's prison ministry program for many years, giving hope to inmates seeking to start over in business after being released. Gary Desler has three sons two of them are owners of businesses, and the other son is a manager and works in the restaurant industry in Maryland. Ed Peters has two daughters that have excelled in science and technology with backgrounds in biochemistry and electrical engineering. I am sure their father's work at Network Solutions in the technology area had an influence on their areas of specialty in the STEM arena.

Even in my own family, my three adult children have started and successfully run their own businesses. Nancy Louise Wheeler is the developer of financial literacy software, Chris White is the owner of his own Fintech company BondCliQ, and Adrian White is the owner of her own web design and marketing company. My older brother's children, Christine White is an attorney with an MBA who provides legal advice and services to newly- formed minority and women-owned businesses in the Atlanta area. Her physician sister, Dr. Nicole Peoples has her own online natural products and health advisory site and blog, and their brother Mark is a software sales manager with Oracle.

There are numerous opportunities for youth to expand their knowledge of STEM through extensive programs sponsored by various federal agencies and private-sector program initiatives. Thousands of kids are being exposed to these areas of technology. They are learning how to apply various types of scientific tools in making decisions. Additionally, they see the technology area as an area for a future career path. I am also seeing a growing number of new companies being started by millennials all over the country. Some are developing new apps to provide shortcuts to new services and products. I am proud to say that during the last couple of years, I have mentored several young men in my church, the Sons of Joshua program, at Kingdom Fellowship AME Church in Silver Spring, Maryland and exposed them to African American inventors and leaders. Additionally, I have worked with other young men who I guided to better job opportunities in the technology area. I am very proud of Ben Harvey and George Sweeper Jr., two young men that I mentored. These are two young men who worked diligently to study and understand the opportunities available in coding and software development. Their individual desire to seek a career in the technology space will benefit them, their families, and their communities.

Even though in the last chapter I focused on African American millennials and their role in the current and next Internet revolution, there is a role for baby boomers who are still active and ready to take on new challenges in their next stage of life. With today's technology, it is even possible for baby boomers to start new careers and identify new job opportunities. However, they first must learn how to use social media platforms to find these opportunities. There are technology tools that provide access to these new job opportunities from their home or office, full or part-time. Recently, I was introduced to two dynamic women helping baby boomers and others find employment opportunities and new careers. Dr. Elisse Wright Barnes and Angela Heath are consultants in the Washington, DC region assisting baby boomers to navigate today's technology. The name of Dr. Barnes' company is Your LinkedIn Driving Instructor, providing guidance on how to access LinkedIn and other social media sites to find job opportunities, answering health and welfare issues, and better understand what your children and grandchildren are doing with the technology. Angela Heath's company is TKC Professional Services, offering training and workshops in helping people navigate the emerging gig economy. Another amazing African American woman providing services to seniors is my wife, Gwendolyn McLaughlin. She is teaching Spanish as a second language in several senior citizen facilities in the Maryland and DC area. Research has found that if seniors learn new things as they age, it prevents or slows down the onset of dementia or Alzheimer's restricting their life possibilities.

Some baby boomers who are proficient with Internet search engines are even finding new companions (husbands and wives), for the next stage of their life. That fact should motivate some of you. I am sure there are other organizations like the ones mentioned above at your churches or senior facilities that offer this type of training. Baby boomers: Do not lose any more time getting connected on the Internet. It opens so many new possibilities and opportunities. People are living longer today; you can learn how on the Internet.

Chapter 20: Corporate and Community Support for Growth and Development

Over the last few years, Jesse Jackson and his Rainbow Push organization has been promoting greater job opportunities at all levels in the tech area, as well as a major advancement in technical assistance and funding programs for minority businesses. Rev. Jackson has been instrumental in addressing the lack of diversity at all levels in many of the technology companies across the country. He has been able to bring attention to the dismal employment record of minorities and women in the tech sector. Some companies have responded with new programs to hire and promote minorities and women. Some have even appointed minorities to their boards which hopefully result in a better understanding of their customers.

Additionally, major corporations are starting to provide capital to the minority community to assist businesses in the technology area. Intel, under the direction of Brian Krzanich, the former CEO of Intel, created a $150 million investment fund for minority technology companies. This fund has been increased to $300 million for investment and job opportunities for minorities and women. Intel was praised by Rev. Jesse Jackson as serving as an example for other major corporations to follow. Some corporations are providing education and training programs within the minority community, as well as business partnering opportunities. Like the late Congressman Parren Mitchell,

Rev. Jackson has been a force by insisting that multibillion-dollar technology companies in Silicon Valley and across the country provide employment opportunities to the minority community.

For more than 50 years AT&T has provided support for the African American and other minority communities through various diversity initiatives. They provided contracting to African American businesses totaling more than $3 billion from both the federal and commercial sectors. Additionally, they provided funding for STEM projects and assistance to historically black colleges and universities (HBCUs). The original Network Solutions benefited from the support of AT&T over the years starting in 1990 by being awarded the largest subcontract awarded to a minority business. During AT&T's s history, they have trained a large number of African American engineers at Bell Labs, the innovative technology unit of AT&T. They also served as a model for other major corporations with vendor diversity programs over the years. AT&T established one of the first Mentor Protégé Programs, creating a partnership with minority-owned businesses in developing joint technologies. Other organizations (Blackrock, JPMorgan Chase, and Goldman Sachs) have established minority venture funds and entrepreneurial training. Goldman Sachs *10,000 Small Businesses* is an investment to help entrepreneurs create jobs and economic opportunities by providing access to education, capital, and business support services. Also, my former employer JPMorgan has established a program to promote entrepreneurship through funding and supporting local community organizations across the country.

Already discussed, several noted minority athletes have recently established their own investment funds and I hope they will be directing some of their money to companies that are minority-owned or in partnership with majority-owned companies with a high level of minority employees. These activities are being fostered by professional athletes and entertainers are long overdue, and we need more to be done today and in the future. The one person that has gone the extra mile in supporting the growth of minority business opportunities in our country is the famed basketball legend, Magic Johnson. He has demonstrated to the majority community that with financial resources we are able to build and grow businesses in the African American community. One of the greatest contributions beyond the millions of dollars he has invested in the minority community is sharing his experience with young people and other businesspeople. I have followed Magic's rise to fame and fortune for a number of years; he is not afraid to take a chance on businesses in our community. From his movie theaters, housing projects, and Starbucks coffee shops, he has brought hope to many.

There are several minority and women tech groups hosting conferences around the country that teach new entrepreneurs how to raise capital from angel investors, VCs, and other investment partners. These programs and information sources did not exist several years ago except for federally funded programs of the U.S. Department of Commerce Minority Business Development Agencies (MBDA) and the Small Business Administration (SBA). The National Minority Supplier Development Council, a privately funded organization, also has provided invaluable assistance to minority businesses across the country. They provide guidance and support to numerous businesses while promoting partnerships with Fortune 500 companies. My local chapter, the Capital Region Minority Development Council, hosts a major annual event to recognize the top 100 minority businesses in the region. The president of this council is Sharon Pinder, a dynamic woman with limitless energy. Sharon Pinder and her staff do a great job of promoting minority business partnerships with corporations and federal agencies. Additionally, Dennis Smith and Kenneth Clark of her

MBDA team have been outstanding at providing minority businesses across the country with sourcing opportunities and capital.

I am also seeing tremendous advancement in the technology area at HBCUs. Over the years, federal and defense agencies have provided contractual support to HBCUs that offered specific skills in selective areas, but generally not in technology. These new programs have been restructured and now focus on providing grants and contracts to HBCUs and colleges with large minority populations in the technology area. There is a nonprofit corporation, MSI STEM Research and Development Consortium (MSRDC) that was established in the Washington, DC area, a couple of years ago and chaired by Richard Hackney, one of my former banking friends. This organization has already prequalified more than 30 predominantly minority institutions to participate in this multimillion-dollar contracting program sponsored by the U.S. Department of Defense. The focus is on identifying colleges and universities that have the capability to provide technology solutions for critical areas related to the war-fighter program, cybersecurity, domestic terrorism, and other critical technology areas. This program will assist these institutions in providing their students and faculty with financial resources and industry experience in the technology area.

Another organization that is providing invaluable support to the minority community is the Bethune-Dubois Institute, located in Silver Spring, Maryland. This organization was established by William Tucker, husband of the late C. Delores Tucker who was the former treasurer of the state of Pennsylvania and major political leader in the Democratic Party. This organization provides financial assistance to minority youth trying to attend college who lack the resources to buy books and other supplies. The organization has expanded its mission by providing training and education in the STEM areas or in other specialties where future employment opportunities exist for young people in our community. I have been a member of the board of this highly successful nonprofit organization that is committed to continuing the legacy of C. Delores Tucker for several years.

In Northern California, there is a nonprofit organization called PACT. This organization has been in existence for more than 40 years helping students and families identify funding sources for college and tutoring services. Their record of accomplishments in this area has been outstanding, assisting thousands of minority students with a college education. One of PACT's other achievements is helping minority-owned businesses and fostering economic development in Northern California. This nonprofit organization has been a role model for other community-based programs throughout the country. My brother, Martin and Louis Barnett have been on their board since its inception, supporting the leadership of Evert Brandon, the head of PACT.

There is a huge gap to be filled in the African American community related to our participation in the technology space. The lack of capital has traditionally been a major issue as well as the need for increased mentoring and support by more experienced businesspeople in our community.

As noted, we are making some progress with closing the wealth gap, but it will take greater participation and support of families, friends, and the community. Additionally, those entertainers, athletes, and businesspeople with the financial resources to invest need to start making a difference and helping other generations to build the next Network Solutions and provide the investment dollars and support to stay in the "game." Today, global technology is moving at lightning speed and will impact every community in the U.S. over the ten years, we will see a tremendous shift in the job market and if we are not prepared, we will be left behind, and most Americans will suffer. The next Internet revolution will displace jobs at all levels; this should be a concern to everyone. Every person in our community will need to have the ability to create a new job by getting in business for yourself or with a group of friends and family.

I hope that my book has provided you with some direction in what we need to do to ensure we are not left at the "station" in phase two of the information superhighway (Internet) in our *Race for the Net*.

Epilogue: Importance of Equal Access and Net Neutrality

During the period of the Clinton-Gore Administration, part of their stimulation plan for the economy was programmed to provide Internet access to all schools in both the urban communities and rural areas throughout the country. They saw the Internet as the economic driver for the U.S. economy that was floundering prior to Clinton taking office. They understood that if the country was to globally compete, providing technology and, specifically, information at all levels in every community, was very important. They established a budget of more than $10 billion to improve the American quality of life and to reinvigorate the economy in the early '90s. Providing access to the Internet was a major part of the program. Here is a summary of the programs that the Clinton-Gore Administration put in place:

1. Increased funds for the National Science Foundation's programs to expand and extend the use of the Internet to educational institutions and laboratories across the country. This included funding to increase the speed of the Internet.

2. Instituted and recommended programs at U.S. Department of Commerce, headed by Ron Brown, the first African American to hold this position, such as using the Internet for telemedicine applications in rural areas of the country.

3. Ron Brown challenged the private sector to develop new business models for low-cost computers and Internet access. He created committees at the Department of Commerce to promote the use of the Internet and to discuss the problem of equal access and the digital divide related to the availability of the Internet in urban and rural communities.

4. The administration mobilized major public and private efforts to bridge the digital divide during his third New Markets Tour when over 400 companies and nonprofit organizations signed a "National Call to Action" to bring digital opportunities to youth, families, and organizations. Education and health programs that could be transmitted over the Internet.

5. Announced in November 1993 by the Clinton Administration under the "Agenda for Action" the federal government's primary role was that of a "traffic cop" ensuring that public rights were protected and that there were universal services available for all.

6. In 1994, only 3 percent of classrooms had computers that were connected to the Internet and only 35 percent had any Internet access at all. The Clinton-Gore Administration, understanding the critical importance of incorporating technology into the classrooms, provided funding through programs like the E-rate and the Technology Literacy Challenge Fund to solve these problems and connect public schools to the Internet. By 1999, there were more than 65 percent of classrooms connected, which rose to 95 percent by the end of the Clinton-Gore term.

7. Vice President Gore played a vital role in securing low-cost connections to the Internet for schools, libraries, rural health clinics, and hospitals with additional funding sources.

Following the Clinton Administration more than 20 years ago, the government continued to protect the rights of the community by ensuring that every community had equal access.

A major issue during this period, and even today, was equal access or net neutrality. There were real concerns during this period from organizations like the NAACP, the Urban League, and the Congressional Black Caucus if the Internet would be available in most urban and rural communities. These organizations observed during the '90s that most of these communities did not have access to the Internet and they questioned why. With an understanding that the future of education and information would be done electronically on the Internet, the question of discrimination started to become a major issue.

During hearings and testimonies on Capitol Hill in the '90s by telecommunications companies and Internet service providers (ISP) they were questioned why the urban communities and rural areas of the country lacked a connection to the Internet. Also, questions were asked by committee members about when they expected these communities to be wired to the Internet. The answers provided by the telecommunications industry were that wiring these communities then and in the near term was too costly. Many residents from these communities due to their income levels would not be able to pay the construction costs to wire these communities with high-speed networks. However, they could use their basic telephone line and a modem (devices connected to the telephone to access outside networks) to connect. This was a much slower option for these communities and limited the availability of educational or health information that could only be transmitted by high-speed networks. I'm sure many of you remember the old modem boxes that were so loud when they tried to connect to the network. Downloading simple files or accessing sites would take forever. Over the years with the deployment of the high-speed network infrastructure, there was no longer any need for the old modem box. However, today there are many rural communities that still utilize some form of modems to access the Internet.

The Clinton Administration tried to address this issue and promote programs like E-rate, in which a fund was set up from fees charged by telecommunications customers to pay for the infrastructure cost of providing services to urban and rural homes and libraries.

Without the support of the E-rate program, there are many schools and libraries in rural areas that would not have high-speed Internet service today.

What Is the E-Rate Program?

On May 7, 1997, the Federal Communication Commission (FCC) adopted Order 97-157 as its plan to implement Section 254 of the 1996 Telecommunications Act. The FCC determined that "telecommunications services, Internet access, and internal connections," including "installation and maintenance," were eligible for discounted rates. Internal connections were defined as "essential element(s) in the transmission of information within the school or library." The level of discount that a school or library received would vary from 20 percent to 90 percent depending on the cost of services and level of poverty as measured by the percentage of students eligible for the national school lunch program. The total amount of money to be disbursed was capped at $2.25 billion or 15 percent.

The FCC designed the application process to promote cost-effective and accountable solutions. As a part of their applications, schools and libraries were required to assess their current technology resources and explain how they utilize them for their educational mission. This assessment had to be certified by an outside organization, preferably the state government. Schools and libraries were required to select vendors through a competitive bidding process publicized through a national website. Record-keeping requirements were instituted to facilitate audits.

The FCC decided to fund E-rate through the same pool of money collected for other Universal Service Fund, or USF, programs. The new language in the Telecommunications Act of 1996 expanded the pool of companies required to contribute. The expanded pool included all companies that provided interstate telecommunications service to the public for a fee. As of 1998, around 3,500 companies contributed to the USF. A company's contribution to the USF is based on its interstate and intrastate revenue from sales to end-users. Companies submit revenue projections, from which the contribution factor is determined and then assessed. This process takes place on a quarterly basis (how the USF works). In order to preserve low-cost local phone service, companies are only permitted to increase interstate revenue to recoup their USF contribution costs.

Modernization

On July 23, 2014, the FCC adopted a broad overhaul of the E-rate program, named the E-Rate Modernization order. The order focused on expanding subsidies for Wi-Fi to a target of $1 billion a year. The move followed a month after a request for reform by President Barack Obama, who had advocated reform of the program during his presidential candidacy in 2007. The move was embraced by many in the telecommunications industry, including AT&T, Comcast, Cisco, and PCIA—the Wireless Infrastructure Association. The reform was also lauded by the American Library Association.

In November 2014, FCC chairman Tom Wheeler proposed the first increase in the E-rate budget, an increase of $1.5 billion. In December 2014, the FCC approved the increase by a vote of three to two, raising the total budget from $2.4 to $3.9 billion.

During President Obama's Administration, they were able to overturn a ruling by the courts on net neutrality. In 2015, under those regulations, broadband service was considered a utility under Title ll of the Communications Act, giving the FCC broad power over Internet providers. The rule prohibited the following:

- Internet service providers could not discriminate against any lawful content by blocking websites or apps.

- Service providers could not slow the transmission of data based on the nature of the content, if it was legal.

- Service providers could not create an Internet fast lane for companies and consumers who pay a premium, and a slow lane for those who did not.

The reason these actions were necessary was due to the importance of the Internet in our society and how no commercial entity should have control of who gets it and who does not. The Internet has become the business highway on which most commercial business is conducted, as well as for small businesses to operate freely. It is the platform that allowed children who are hospitalized in small towns or rural areas to still get an education.

In the 25 years since Network Solutions opened the door to providing the global Internet to the public, it has become a tool of necessity for everyone. It has become more important than the TV in the home and almost as important as the telephone. In fact, the telephone and the TV all use services provided by the Internet today. There are millions of people throughout the U.S. who are dependent on Internet service for so many applications.

I believe telecommunications companies will provide services at a reasonable price and access to everyone. However, having control over who gets the Internet or not when it comes to the bottom line will be a difficult decision.

One of the lessons we have learned from the opening of the cable industry and the FCC approval to merge and reduce competition is that they promised a better rate structure. However, what we received were substantially higher costs even though they said it would not happen. In some cases, cable rates were increased by more than 100 percent within one year after allowing for a competitive market.

However, cable access is not the same as the Internet, even though many providers today use the Internet to provide communications and information to their viewers.

The problem for the African American community with the recent lifting of the net neutrality rule allowing the telecommunications providers to control the price and access in our community for major users of social media like Facebook, and Twitter, etc. So, if there is a rate increase for social media sites, it will have a major economic impact on our community. This also will be true for video games, for which we are major users.

This is more of a reason why we need to participate in this Internet economy as more than consumers. With just being a consumer, you have very little control over your future and how to increase your level of wealth. We can complain about the loss of net neutrality, but it will not change the situation.

The result of the impact of eliminating net neutrality is the ISPs or communications companies' sensitivity to the importance of the Internet to our society, regardless of economic status. I hope that the price for high-speed networks will not change much over the next several years while technology advancement lowers the unit cost.

Many of the social and political organizations representing minority and rural communities in the country believed that the Internet would not pass through their towns or cities. This would result in the traditional separation of the benefit of the Internet between the haves and the have-nots. Let me give you an example of the problem in a simple analogy. Let's say that color TV was only available today to those areas of the country where wealthy people live and the rest of the country (more than 75 percent of the population) could only receive a black and white transmission. Those individuals and families with color TVs would be able to recognize the colors of opposing teams on the football field or basketball court. Those receiving color transmission would also be able to distinguish more details on educational channels and learn more about science and technology. This problem does not exist today since we all have access to color TV.

Since access to the Internet provides so many services that benefit children, families, and organizations, restricting access and service to minority and rural areas of the country would cripple a segment of our economy.

The penetration rate is less than 15% if your income is less than $25,000. States like Mississippi still do not have Internet access in most rural areas. The rural areas of Kentucky, West Virginia, and New Mexico are also similarly affected.

The recent FCC dropping of the net neutrality regulation and funding support were discriminatory against the poor and disenfranchised people in our society, mainly people of color.

The benefit of the Internet can be compared to the telephone, electricity, and water. It is necessary for everyday life and needs to be protected like other utilities. As the Internet gains greater access to our everyday life, it has become a necessity for learning, health, and a sense of well-being. Earlier I mentioned how the Internet has saved lives during disasters and provided mechanisms for generous donations so people could rebuild their lives. There are more than 21 million people today across the U.S. that need and require broadband access to enable them to have normal and productive lives. Society needs to make sure that these communities are not separated from the rest of the world due to the lack of current technology. We must end the digital divide in our country forever.

References

Feifer, Jason. "Take the Risk Editor's Notes." *Entrepreneur Magazine,* Nov. 2017.

Jayson, Sharon. "More Than a Third of New Marriages Start Online." *U.S. Today,* June 2013.

White, Albert. "A Commitment to Mastery." *Network Solution Marketing Dept.,* 1992.

Achenbach, Joel. "Disaster Clams Soars in Years Calamities." *The Washington Post,* Nov. 2017.

Adams, Genetta M. "17 Black Internet Pioneers." *The Root,* Feb. 2012.

Adams, Janet, and Paul Overberg. "Millennials Near Middle Age in Crisis." *Wall Street Journal,* May 2019.

Anderson, Monica, and Aaron Smith. "5 Facts About Online Dating." *Pew Research Center,* Feb. 2016.

Anderson, Monica. "6 Key Findings on How American See the Rise of Automation." *Pew Research Center,* 4 Oct. 2017.

"AT&T Awards Its Largest Contract to Minority Subcontractor Ever." *Banner News Paper,* Sept. 1990.

Barmore, Jasmin. "Sharing the Joy of Eating with Faithful Fans Online." *New York Times,* June 2019.

Bell, Tracy. "Michael Daniel Discussion New Book: The Monetization of the Internet." Feb. 2014, p. 1.

Beyster, Robert, and Mike Daniel. "Names, Numbers and Network Solution." *The Monetization and of the Internet*, 2013.

Bigelow, Bruce V. "The Untold Story of SAIC, Network Solutions, and the Rise of the Web." *Xconomy*, July 2009.

"Black Millennials Close the Digital Divide." *Nielsen 2016 Report*, Oct. 2016.

Bowen, Fred. "Jackie Robinson: A Towering Figure in American History." *Washington Post*, Apr. 2018.

Brown, Eliot. "Tech First Muscle into Each Other's Turf." *Wall Street Journal*, Oct. 2017.

Brustein, Joshua. "White Male Venture Capitalists Tend to Fund White Male Founders." *Bloomberg Business*, May 2018.

Burns, Lawrence D. "Late to the Driverless Revolution." *Wall Street Journal*, 18 Aug. 2018.

Cohen, Noam. "The Players Ball." *New York Times*, May 2019.

Coleman, Lauren Delisa. "Black Founders of Internet Domain Registry Network Solutions, Reminisces on Racial Barriers in Tech." *The Grio*, June 2012.

"Congressman Parren Mitchell." *Baltimore City Paper*, June 2002.

"Congressman Parren Mitchell." *Encyclopedia.com*, 2019, encyclopedia .com/Parren Mitchell Congressman Parren Mitchell.

"Cooperative Agreement Between NSI and U.S. Government." *ICNN*, Jan. 1993.

Cross, Tim. "Human Obsolescence." *The World in 2018*, 2018.

"Domain Names Services No Longer Subsidized by Taxpayers." *National Science Foundation*, Where Discovery Begins, 14 Sept. 1995.

Fraser, George. "How African American Lost Control of the Internet: Did Race Matter?" Mar. 2000.

Friedman, Thomas. "2007 The Year Most Technology Was Discovered." *Thanks for Being Late*, Nov. 2016.

Glaser, April. "The U.S. Will Be Hit Worse by Job Automation Than Other Major Economies." *Recode*, Mar. 2017.

Guynn, Jessica. "Jackson to Tech: Double down on Diversity." *U.S. Today*, Apr. 2018.

Haddon, Heather. "To Speed Up, McDonalds Enlist Robots." *U.S. Today*, June 2019.

Henderson, Gina. "Net Loss (How A $21 Billion Internet Deal Got Away)." *Emerge Magazine*, May 2000, pp. 37–38.

"History of National Science Foundation." *National Science Foundation*, 2019, Wikipedia.org/wikiNationalal.

"How the Internet Has Changed Dating." *The Economist*.

"Job Lost, Job Gained, Workforce Transitions in a Time of Automation." *McKinsey Global Institute,* Dec. 2017.

"Job Openings and Labor Turnover March' 18." *Bureau of Labor Statistics News Release*, Mar. 2018.

Kapner, Suzanna. "Sears: How It Lost American Shoppers." *Wall Street Journal*, 16 Mar. 2019.

Kinnon, John. "FCC Reverses Rules on Net Access." *Wall Street Journal*, 15 Dec. 2017.

Lankford, Ronnie D., and Micah Issit. "Parren Mitchell Biography." Jan. 2008.

Lebowitz, Shana. "18 Habits of Highly Successful People." *Business Insider*, Oct. 2016.

Leiner, Barry, et al. "History of the Internet." *Internet Hall of Fame*, 1997.

Merle, Andrew. "The Reading Habits of Ultra-Successful People." *Healthy Living*, Apr. 2016.

Mims, Christopher. "May I Please Disrupt Now?" *Wall Street Journal*, 21 July 2018.

Mitchell, Josh, and Andrea Fuller. "Student Debt Crisis Hits Hardest at Black Colleges. April 2018.

"Network Information Services Manager Award Abstract #9218742." *Where Discoveries Begin*, Dec. 1992.

Pegoraro, Rob. "Wrong on the Internet." *The Washington Post*, March 2017.

Perrin, Andrew. "Smartphones Help Blacks, Hispanics Bridge Some - but Not All -Digital Gaps with Whites." *Pew Research Center*, 31 Aug. 2017.

"Report Job Sport Suggests 38% Will Be Automated by the Early 2030s." *Daily Mail*. March 2017.

Schreiner, Taylor. "Online Shopping Vs Brick and Mortar." *U.S. Today Money Section,* 24 Apr. 2019.

Spenser, Michael K. "Why Capitalism Isn't Working for Millennials." *Utopia Press,* Mar. 2016.

Stanley, Morgan. "Growth of the Internet in the 90s." 1995.

"The Global Information Infrastructure Agenda for Cooperation." *National Telecommunications and Information Administration,* June 1995.

Turner, Nathaniel. "MBELDEF Profile Parren Mitchell." 2019.

Warner, Melanie. "Pete Musser Built His Company over 40 Years. Then He Was Seduced by the Internet. and His Billion-Dollar Fortune Melted Away." *Fortune Magazine,* 5 Mar. 2001.

Washington, Technology. "RMS Moves Beyond Federal Sales." Sept. 1996.

Wasow, Omar. "Biography." 2019, http://www.omarwasow.com/.

"What the Future of Work Will Mean for Jobs Skills and Wages." *McKinsey & Company,* 11AD.

White, Albert. "The Information Superhighway: Will It Be a Boom or Bust for Minority Business."

White, Albert. "At the Junction." *Minority Business Enterprise,* 1994.

"Who Will Care for Our Seniors?" *New American Economy,* Sept. 2016.

"Network Information Services Manager Award Abstract #9218742." *Where Discoveries Begin,* Dec. 1992.

Stanley, Morgan. "Growth of the Internet in the 90s." 1995.

"The Global Information Infrastructure Agenda for Cooperation." *National Telecommunications and Information Administration*, June 1995.

Turner, Nathaniel. "MBELDEF Profile Parren Mitchell." 2019.

Washington, Technology. "RMS Moves Beyond Federal Sales." Sept. 1996.

Wasow, Omar. "Biography." 2019, http://www.omarwasow.com/.

White, Albert. "The Information Superhighway: Will It Be a Boom or Bust for Minority Business."

White, Albert. "At the Junction." *Minority Business Enterprise*, 1994.

"Who Will Care for Our Seniors?" *New American Economy*, Sept. 2016.

Names Mentioned in Book and Page Numbers

About the Author

Albert E. White

Albert White has over thirty years of business and finance experience as an advisor to some of the most successful CEOs in the U.S. His vast experience is in the technology areas of Health Care, Energy, Disaster Services, Communications, and Internet of Things (IoT).

He is a known visionary in identifying market niches.

Mr. White has been successful in building capacity for small and medium-sized federal contractors. In the 1990s, he was responsible for negotiating the largest subcontract that AT&T had ever signed with a Minority-Owned Business. He achieved this result as a member of the original management team of Network Solutions, the company that first commercialized the Internet. He was responsible for creating the company's marketing strategy to promote the initial use of the Global Internet to the public.

For more than sixteen years, Mr. White has been an advisor (and former CEO) for Dr. Lonnie Johnson, one of the leading and most successful inventors in our country. Dr. Johnson has more than 100 patents, including the Super Soaker and Nerf Gun. Mr. White negotiated a contract modification with Microsoft for Dr. Johnson valued at $36 million.

Mr. White has an extensive finance and investment background. Early in his career, he was an International Banking Officer with JP Morgan and subsequently Bankers Trust Company (acquired by Deutsche Bank). He later served as Senior Consultant to Safeguard Scientific, a leading publicly-owned venture capital company.

In 2016, Mr. White formed a consulting service, Polaris Advisory Group LLC, to provide funding and marketing strategies for Minority and Women-Owned Businesses with a focus on Energy, Health Care, Information Technology, and Government Contracting.

As a respected authority, Mr. White has been featured in the New York Times, Black Enterprise Magazine, New York Daily News, and other publications. He is a noted presenter at conferences sponsored by the Bank of America/Merrill Lynch and the Rainbow Push Coalition. His presentations are on two topics: Raising Capital for International Projects; and How to Identify Business Opportunities in the Future for Minority Firms.

Mr. White earned his MBA in Finance from Columbia University and a BS in Marketing/Economics from the University of Denver. He did postgraduate work at the University of Michigan in Marketing Strategy.

He has held board seats on publicly owned and non-profit companies.

A native of Brooklyn, New York, Mr. White resides in Silver Spring, Maryland. He has three adult children, who all operate their own companies.